A SUMERIAN OBSERVATION
OF THE KÖFELS' IMPACT EVENT

A SUMERIAN OBSERVATION OF THE KÖFELS' IMPACT EVENT

a monograph by

ALAN BOND

and

MARK HEMPSELL

A Sumerian Observation of the Köfels' Impact Event

Alan Bond
Reaction Engines Limited,
Culham Science Centre, Abingdon, Oxon OX14 3DB, UK

Mark Hempsell
University of Bristol,
Queens Building, University Walk, Bristol BS8 1TR, UK
mark.hempsell@bristol.ac.uk

ISBN 1904623646

Revised 2008

First published in Great Britain (2008) by Alcuin Academics
Typeset by e-BookServices.com

CONTENTS

LIST OF FIGURES

LIST OF TABLES

PREFACE

For several years we have been conducting research into the possibility of a major Near Earth Object impact in the Early Bronze Age. While this was not in itself an original hypothesis, our interest was in exploring whether Köfels in Austria could have been the site for such an event which, to the best of our knowledge, was a conjecture that had not been considered before. When we started the project this premise was highly speculative and the work was done for our own intellectual stimulation rather than in any expectation of reaching any conclusive, let alone publishable, results. However the tone of the project changed when our attention was drawn to tablet K8538 (commonly called the "Planisphere"), a cuneiform text in the British Museum. There has never been a comprehensive and consistent translation of this unique tablet, but it had several aspects that led us to the conclusion that it might relate to a Near Earth Object impact and on the basis of this impression we undertook to examine it in detail.

We had two advantages over previous researchers who have tackled K8538. The first advantage was that we had a clear context regarding what the tablet might be about. Context is always important when dealing with Sumerian cuneiform due to the high level of ambiguity compared with modern written languages, but it is especially important in the case of this tablet where its purpose seems to have been more as an "aide memoire" to the writer than a means of communication with other people.

It is therefore important to know that a Near Earth Object impact is an event that might have been observed. It also helps to have some experience of astronomy and understand the sort of factors that affect what could be seen and what would be of interest to an observer. We are certain that modern astronomers who follow the detail of the tablet will feel an empathy with a skilled and objective brother astronomer - even after 5000 years. Without this background knowledge understanding K8538 would be very difficult.

The second advantage we had was new software tools that were not available to any previous researchers. The programs used for trajectory modelling were non-commercial; they were modified versions of programs that had been used on a previous project. The other programs, used to model the position of stars and planets, were widely available commercial packages. When these programs were combined we could directly compare the tablet's pictures to the actual sky. Without this ability to "see again" what the Sumerian astronomer saw, we believe understanding the tablet would be impossible and we believe this was the key problem that has prevented previous researchers fully understanding K8538.

As a result of this examination we believe our original speculation regarding the tablet was correct, that is the tablet is an astronomer's contemporary record of a kilometre class Aten asteroid as it approached Earth to impact at Köfels in Austria. What we were completely unprepared for was the objectivity and accuracy of the observation, which gives an incredibly rich insight into the event it describes. Clearly these conclusions merited publication; hence this monograph.

We decided to report our results in a monograph for two reasons. The first reason is that the arguments are complex. There are three major interlocking areas, the tablet itself, the Köfels site, and the heliocentric orbit of the object. Each of these areas is complex in its own right, yet each is internally consistent. Then the stories they each tell individually are completely (and very precisely) consistent with each other. Thus we have jigsaws that are themselves pieces within an overall jigsaw and the proof lies in that the end result is such a clear unified picture. An argument this complex cannot be fully outlined within the word constraints of a journal paper whereas in a monograph we can both present the full arguments, and cover some of the secondary material, needed to establish the veracity of our contentions.

The second reason for using a monograph to report this work is that it is highly multidisciplinary in nature and this creates problems for a conventional journal paper. Academic journals are focused on specific subject areas with specialist editorial boards, and specialist referees to review the papers submitted. Not surprisingly, editors and referees have difficultly handling papers that have substantial content outside the journal's field and yet which form an essential part of the argument. This problem then extends to the readership of a journal, which is also restricted to specialists in one subject rather than the totality of people who would have an interest in the work. A book enables us to engage all the potentially interested communities on an equal basis, even if we have lost the advantages of peer review.

Of course a project such as this has meant we are greatly indebted to many people for the help and support they have given us. In this regard we are greatly obliged to the British Museum for their assistance with our project, especially Christopher Walker and Jon Taylor of the Department of the Ancient Near East. Visitors to the Museum's excellent displays perhaps do not realise the degree to which the Museum provides a support service to scholarship and research, and certainly we found their considerable help invaluable.

We also need to thank the Department of Archaeology at the University of Bristol for their advice, support and the provision of very helpful research seminars, and we especially would like to record the special interest, encouragement and help of Volker Heyd, Lecturer in Archaeology.

We would like to thank Flora Turner of the Croatian embassy and Nikša Petric for translations and other help related to the Hvar pottery. And we would like to thank Natalie Allred and Colin Hawkins for various translations of German texts.

There were also a great many people - colleagues, families and friends - who have furnished a great deal of useful general advice and support for the work. There are too many to name them all but the following (in alphabetical order) have all merited our special thanks for their "above and beyond the call of duty" actions or advice: Don Carleton, Diane Holmes, Friedrich Horz, Chris Rose and Carrie Wattling.

While we wish to acknowledge their help, we must also add the disclaimer that just because these wonderful people helped us, this does not necessarily mean they endorse our conclusions, and they are certainly not responsible for any mistakes we may have made.

Finally we must also acknowledge the unknown Sumerian astronomer who originated K8538 and thus has provided one of mankind's most important scientific records. Out of respect we have personalised him (or maybe her?) with the name LUGALANSHEIGIBAR *"great man who observes the heavens"*.

Alan Bond
Mark Hempsell
May 2007

1

INTRODUCTION

This monograph considers two long standing enigmas; a strange geological feature centred on Köfels in the Austrian Tyrol and an unusual cuneiform tablet in the British Museum collection which is commonly called "the Planisphere". There has been considerable past work and speculation on both subjects that stretch over more than a century but, despite this, neither puzzle has a conclusive or generally accepted explanation. The argument presented in this monograph is that the tablet is a contemporary Sumerian observation of the exo-atmospheric passage of a kilometre class asteroid that later impacted at Köfels creating the features found there.

The structure of the argument is that, using the Köfels' impact as context, the tablet can be translated and its contents understood. From the detailed observation contained in the tablet a trajectory of the object can be determined, which when tracked back to a heliocentric orbit, is found to be an Aten class asteroid in a resonant orbit with the Earth. When the same trajectory is tracked to the ground at Köfels, a series of complex interactions are revealed that explain why the site does not have an impact crater some kilometres in diameter, which is what would normally be expected from such a large object.

Thus the overall argument presented here has a circularity to it, in that the Köfels impact is a context presumption in the interpretation of the Planisphere tablet, then the content of the tablet supports the interpretation of the Köfels site. It is the complete consistency of these interpretations to a very high precision, supported by the derived resonant heliocentric orbit, that provides the proof these interpretations are correct. But the keystone in the argument is the tablet so we must start by examining what is known of its provenance and history.

The Planisphere tablet [1] (Figure 1) is number K8538 in the British Museum cuneiform collection. It is presumed to have been found in the royal library in Nineveh by Henry Austin Layard in the mid nineteenth century. However site reporting at the time was not to the standards we expect of modern archaeology and the tablet's discovery is not recorded, so it is not possible to determine conclusively where or when it was found.

The tablet is an unusual text in many ways. It is a disk, about 18 cm in diameter, and has a marked border; both are rare features. It has text on only one side, the other being unmarked and a slightly convex bowl shape, a shape that is believed to be unique. The text and diagrams are divided into eight equal sectors. The tablet was not found complete; about 40% is missing and these areas have been restored in plaster as blank. The damaged region was caused by fire and occurred in the Assyrian period. Consequently less than half the tablet can be clearly seen, but even so a surprising amount of the information remains intelligible and it uses some

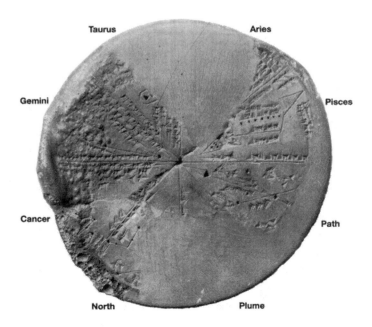

Figure 1:   Tablet K8538 (British Museum photo B580)

62 cuneiform sign variants, about 10% of the total known for this written language.

The subject matter is clearly astronomical as it contains pictorial representation of constellations (another unique feature) and identifiable constellation names. It was first studied by Bosanquet and Sayce [2] who, like many later researchers, considered it to be a compete sky survey leading to its later dubbing as the "Planisphere", although no researcher has managed to show that it is a planisphere in the modern sense of the word.

The archaeological context and the cuneiform style date the tablet to the Seventh Century BC, but this is of limited value in interpreting its content. Translating and interpreting cuneiform is beset by the many ambiguities inherent in this early writing system. A text of this period may be written in either the contemporary Akkadian language or the Sumerian language, used for scholarly and business purposes long after the language had died as a spoken tongue. Further it is known that when scribes copied earlier work that used older variations of cuneiform, although they would be pedantically accurate to the point of absurdity on every other aspect, they would update the cuneiform to the latest version. Therefore although the Planisphere is written in Sumerian, in a late cuneiform style, it could be either a contemporary Assyrian work or a copy of an earlier work from the Bronze Age Babylonian or Sumerian civilizations.

When consideration is given to these uncertainties and the ambiguities caused by homophony and polyphony in early Mesopotamian writing, and that these problems are compounded by the abbreviated note-like style on the tablet; one must conclude it is almost certainly impossible to obtain a full translation or interpretation of the information it contains without a clear idea of the context. However, even without this context, some conclusions can be reached and previous researchers [2,3,4,5] have reached the current understanding that the subject matter is astronomical and some constellations are clearly identifiable by names known from other texts from the region, especially the MUL APIN [6,7].

The most comprehensive work on the tablet to date is by Koch [5]. While he believed he understood which region of the sky each sector referred to, he was unable to explain the tablet's purpose. This work does have some problems, such as the use of stars below visual magnitude in the constellation reconstructions and, while the overall identification is plausible, it is not conclusively proved.

Following all modern researchers, this translation of K8538 has been made possible by the drawing produced by King of the British Museum (Figure 2) in 1912 [3]. Although King had little idea as to the tablet's subject matter, close inspection and comparison between the drawing and the tablet has not revealed any cuneiform logograms that required subsequent reinterpretation, even in the heavily damaged areas. Therefore King's cuneiform identifications have been used as the basis of the translation from Sumerian.

However, King's copying of the diagrams is less accurate and this could be a source of confusion as in some cases King's random copying errors have ended up producing a better match to the actual sky than the original tablet. However this is clearly a result of coincidence and King's drawings have been ignored in this interpretation. The original tablet has been used as the sole source when comparing the diagrams with the actual night sky.

There has not been a consistent and established sector labelling of the tablet by past researchers and none of the past labelling relates

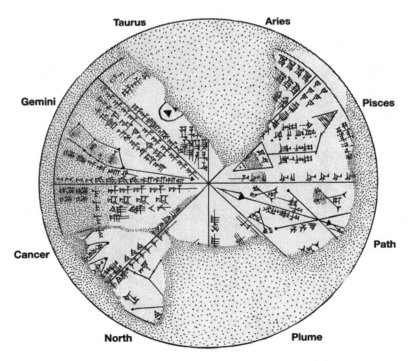

Figure 2: King's Drawing of K8538 (British Museum [3])

well to the tablet's contents, so a new labelling has been used in this translation and this has been added to Figures 1 and 2. The labels were derived from the sector contents as determined by the translation.

What this translation establishes is that, although the cuneiform style and the presumed context both date the tablet to the Assyrian period at around 700 BC, it is in fact a copy of a much older work that would have used Fourth Millennium BC cuneiform.

We can be certain that the tablet is not a First Millennium BC work from the position shown for the celestial equator in the Pisces Sector, which is only correct for a period between 4000 BC and 2000 BC; that is the Early Bronze Age when Sumer existed as a separate culture. This opens an intriguing possibility: if the copier understood the astronomy on the tablet he must have known about the precession of the equinoxes more than half a millennium before Hipparchus, who is credited with the discovery.

Almost nothing is known about Sumerian astronomy [8] except that the extensive use of Sumerian names for celestial objects by the later Semitic Mesopotamian cultures suggest they learnt their astronomy from the Sumerians, in the same way as the use of Latin and Arabic names and terms indicates the sources of modern astronomy. From this we can deduce that Sumerian astronomy must have had some merit. This is confirmed by K8538, which is the first real insight into Sumerian science and it proves that they did indeed possess a very sophisticated and advanced understanding of the heavens.

The tablet describes a summer night sky in 3123 BC shortly before dawn. The emphasis is on recording transitory features such as the position of planets, clouds and, more spectacularly, the path of a large incoming asteroid. The obvious conclusion from the contents is that it is a "field record" made by Sumerian astronomers at an observatory site.

Although there are no other known examples of such field observations, the astronomical tablets containing detailed records that have been found in Mesopotamia prove such field observations must have been made. It is not credible to expect astronomers to simply remember such details for later recording. K8538 is believed to give us an insight into how these observations were made, at least by the Early Bronze Age Sumerian astronomers.

With this interpretation two unusual features of K8538 can now be explained. The first feature is the border clearly (and it is believed correctly) shown in King's drawing. This is unusual in cuneiform tablets, which rarely leave a gap between the edge and the text, let alone draw any margin line. The second feature is the reverse of the tablet, which is not flat, but shaped like a smooth convex shallow bowl. It is a very convenient and comfortable shape to hold in one hand. Both these features would be explained if the original was a shallow pottery bowl filled with soft clay. This would be a more convenient writing surface than a simple slab of clay, given it is used away from the scriptorium, while using astronomical instruments and at night. It is in essence a notebook intended for temporary records and once the results were permanently recorded it was presumably wiped clean for reuse.

This speculation would also explain why no more of these field notebooks have been found. Firstly, although records would be made every night and possibly more than one a night, they would not last long, a week to a month but seldom longer. As they were then wiped clean for reuse the number of these writing bowls would be small. Further they were probably only rarely in "libraries" or "scriptoriums" spending most of their working life in the observatory. It will be shown in the Gemini sector that the tablet suggests the observatory was some way from its related city.

The question then arises if these nightly records were temporary, and were normally destroyed, why did this one survive (either the original or as copies) for over two thousand years to be available for copying by the Assyrians. The answer possibly lies in the extraordinary event it records. The impact of a large asteroid at Köfels would have had a profound effect on all the societies that witnessed it. It seems very likely this event gave rise to the Sumerian story that the god ENLIL took control of the Universe from the sky god AN and the earth mother goddess KI, an event the Sumerians considered to be historical [9]. If the hypothesis that the tablet is associated with this myth is correct, then the religious implications of the observation recorded mean that it would have very special significance and would explain why the tablet was preserved.

The translation shows that the tablet is composed of short notes. Where constellations are identified, and in particularly where they are drawn as diagrams, it is in order to locate the position of transitory phenomena. The only inconsistency with the tablet as an observational record is that the names of the four planets seem to be missing. It may be because the night on night recording made this unnecessary. However this is thought an unlikely practice, given that one of the planets involved is Mercury, which is notoriously tricky to observe.

An alternative explanation as to why we do not see the planet names may be due a combination of reasons. The planet in Taurus may be named in the missing section of that sector. The planet in Gemini may be named as Saturn but in an archaic form we do not recognize. The two planets in Cancer may not be named because, before the names were added, the observer was distracted by the Köfels asteroid, and never got around to adding the names later.

The tablet's eight sectors seem to be little more than the equivalent of notebook pages. Each sector is generally (but not always) devoted to one constellation of the zodiac and the region around it. However, the author seems willing to overlap, expand out to wider regions and to focus in on detail, so does not seem to have any rigid rules about sector layout, scale or content. The orientation of the cuneiform logograms and the diagrams show sectors which should be read when they are on the left, with the lower line roughly indicating down, but only in one case (the Cancer sector) does the lower line strictly represent the horizon, when it is labelled as such.

It is appreciated that the derived date of 3123 BC raises an issue as to whether written language had developed sufficiently to produce a document as sophisticated as K8538. The first counter argument to this is that the archaeological material upon which we base our understanding of the development of writing is sparse and cannot be regarded as conclusive, and the Fourth Millennium date indicates that the current perception of the state of written language of the time may require revision. Further some cuneiform scholars argue early Sumerian writing was as fully developed (or nearly so) as modern writing [10].

However it must also be pointed out that from a language point of view K8538 is not as sophisticated as it may at first appear. All the text is in the form of very short notes; there is no extended prose or abstract argument. This may simply be because it is the style appropriate for its notebook function, but it is also consistent with a written language not yet capable of much complexity of expression.

Another issue arises if the original tablet was written in the earliest cuneiform style. The style used in the late Fourth Millennium has a pictogram form and if applied to K8538 would produce a result much more like a picture than a written text. For example the cuneiform sign for AN (Borger's index 13 [11]), in this context translated as "heaven/sky", is used in a repeated pattern in the Cancer and Gemini sectors to indicate a clear sky. The original Fourth Millennium pictogram for this was a drawing of an asterisk-like pointed star that would give these sectors an even more picture-like appearance and the original author may never have intended these pictures to be read as words.

2

ANALYSIS METHODS AND ASSUMPTIONS

2.1 CONTEXT ASSUMPTIONS

Ambiguities present in cuneiform writing mean that the context of the subject often needs to be well known before the text can be properly understood. Despite the diagrams it has still proved difficult for past researchers to make sense of Tablet K8538 without knowing the context. This translation used four basic working assumptions regarding the tablet's context. The first three assumptions were originally derived from earlier work on the Köfels event and although simply stated here, they were in fact the outcome of a long iterative process of proposition and disproof. That is, many alternatives have been considered and this context is the one that was found to produce the most consistent result, although a few aspects of the tablet remain a puzzle.

The first assumption is that the tablet is an astronomer's record of a particular sky at an instant of time and therefore shows transitory features including the areas obscured by cloud. Given it is written in cuneiform and was found in Mesopotamia, the observing latitude should be consistent with 30°–37° latitude and from a date when cuneiform was in use.

The second assumption is the interpretation of a line of cuneiform in the Cancer sector as a time and date, giving the day of the lunar month and time through the night from the last sunset. An adjacent line is taken as the time remaining to sunrise. Ambiguities in written Sumerian numbers make the initial adoption of these values an 'act of faith' but this is justified retrospectively by the consistent picture which emerges for the astronomy portrayed on the rest of the tablet when compared with computer simulations.

The third (and key) assumption is that Tablet K8538 records an observation of an Aten class asteroid that impacted the Austrian Alps at the locality of Köfels at the end of the Fourth Millennium BC. The study of Tablet K8538 followed more general work on the Köfels event that had produced a trajectory, time of year and an early Bronze Age date and this was available before work was started on K8538. In hindsight this earlier work, while not perfect, proved to be remarkably accurate, and may have involved an element of luck. However without this context much of the detailed understanding of this tablet would not have been possible

The fourth assumption is that the subject matter of K8538 is not mystical. It is believed to be an objective record of an observation (what we would now call scientific) and as such a physical interpretation has been selected for words that may have a theological interpretation as well. Thus for example, for Borger's sign number B13, AN "sky" is chosen consistently over the alternative DINGIR "deity" and this is found to give a consistent interpretation of the text.

To obtain a translation that is fully consistent with this proposition there are a few places where it departs from strict adherence to the modern lexicons. These changes are: either using homophones to provide alternate meanings, or a symbol has been used to signify a new word or meaning. These departures from normal translation assumptions are discussed on a case by case basis during the translation.

These translation assumptions mean that by the strictest standards of proof this fourth assumption is not conclusively verified. However there are no cases where these assumptions drastically alter the overall conclusions about the tablet contents; it is simply that without them much more of the text remains enigmatic.

2.2 SKY MODELLING

In order to use these assumptions to understand the tablet it is necessary to have a detailed reconstruction of the night sky in question. This study used the commercially available Redshift night sky modelling program published by Maris Multimedia / Focus Multimedia Ltd. This programme can reproduce the night sky from any location on earth for any time from Julian date 0 (1st January 4713 BC) including planetary positions, proper motion of the stars and changes due to the procession of the equinoxes. Without this sort of astronomical programme, which can quickly reproduce the sky as seen by the observer, it is unlikely that tablet K8538 could be fully interpreted. Such a tool was not available to any previous researchers and explains the difficulties they clearly had in understanding the tablet.

The project has used both Versions 4 and 5 of Redshift and no discernible difference as to star or planet positions was found. Inaccuracies in the earlier versions of this program were known and a "sanity" test looking at early solar eclipses confirmed the known problems have been resolved in the program versions used on this study.

As a further check on the results given by Redshift, a comparison test was performed with another commercially available night sky modelling program – Starry Night (Enthusiast version 4) published by Imaginova Canada Ltd. This confirmed the star and planet positions for the key dates considered by the project. We presume this is due to the use of common modelling and data sources which are judged to be the best available.

It is known that there are uncertainties in the planetary and stellar positions over the 5000 years. But generally the close match with the tablet confirms generally that these are not significant to the project with one possible exception (Saturn in Gemini). The planet ephemeredes are derived from VSOP87 which is developed by the Bureau des Longitudes (Paris) [12] and is quoted as giving a typical accuracy of 1 arc minute over thousands of years. Saturn is the least accurate but over 5000 years the error is believed to be less than a quarter a degree.

A feature of Redshift is a crude model of the affect of dawn light on star visibility. These results were confirmed by a more refined analysis. This suggests that stars below magnitude 3 would not have been visible at the exact time stated, but the lighting would have been changing rapidly and stars down to magnitude 4 would probably have been visible at the time inscription of the tablet was started. It will be shown how this affected the way the constellations were represented.

2.3 THE MUL APIN

The general layout of the Mesopotamian heavens is known primarily from a cuneiform text called the MUL APIN (from its first words MUL APIN, Constellation of the Plough). Two modern interpretations of the MUL APIN have been used for this work; Hunger and Pingree [6] and Papke [7].

The MUL APIN text has no diagrams, it takes the form of lists of constellations, important stars and other points, in tracks across the sky. Modern researchers have attempted to match the constellations named to the night sky. Hunger and Pingree and Papke agree the general layout but differ in detail. The comparison of the MUL APIN and K8538 confirms this general layout. However at two points the constellation diagrams show that the actual constellation differs from both of the modern interpretations. Fortunately one of the two labelled constellations shown is the MUL APIN itself, which proves to be part of modern Pisces, so there is scope for a further re-interpretation of the MUL APIN in light of the additional insight K8538 provides.

Despite the problems of minor discrepancies, the MUL APIN text has often proved useful in interpreting some aspects of K8538. However at other times it is confusing and disjointed and with some evidence (such as the interpretation of the significance of "star of abundance through persistence" and the inclusion of useless helical rising data), which suggests the author (or more accurately compiler) did not fully understand what he was writing about. It is suggested here that the MUL APIN text is a compendium of texts, only half understood by their compiler, which were already very ancient (over a thousand years) at the time of their compilation (a similar conclusion has been reached by Hunger and Pingree [6]).

2.4 TRANSLATION METHODOLOGY

The task of searching all of the possible homophones and polyphones is tedious and if all of these were presented and sifted logically here, this monograph would be unmanageable. In the following interpretation only the word that gives the most credible meaning has been used and the Borger index [11] for the logogram has been written alongside. This translation has used Labat [13] and Lassine [14] for Cuneiform symbol identification and has used Labat [13] and Halloran [15] as the main Sumerian lexicon sources. If doubt exists, or credible alternative meanings have been found, they are discussed with the preferred meaning. Damaged signs that are not decipherable are written %%%.

In the following sections a photograph of the sector in question is accompanied by King's drawing of that sector and a separate diagram gives the Borger index for each sign (in the text these are prefixed with a B e.g. B25). Each sector is read with the sector horizontal, and with the periphery of the tablet to the left and the tablet centre to the right. Simulations of the sky for the preferred date of 29th June 3123 BC are presented to illustrate the quality of the correspondence achieved for this date. Although some other dates are also possible candidates with different planets, the overall interpretation of the tablet's purpose is not affected by the choice of date.

The translation follows the convention that Sumerian words in an Akkadian document are shown in capitals, such as the title of the MUL APIN. Although the original was clearly written entirely in Sumerian, in this respect the copy K8538 has been treated as an Akkadian document, and therefore all text from it has been shown in capitals. Having prepared the manuscript using this convention, it was seen that the few remaining uses of Sumerian words were less clear and so it was decided to extend the convention and all Sumerian words are shown in capitals.

This translation starts in the Cancer sector and then works through the sectors in clockwise direction (and increasing azimuth) through to the North sector, which completes a picture of the entire sky.

The selection of a starting sector for the following presentation is somewhat arbitrary although it is possible to make an intelligent guess as to the order in which the sectors were written. The first three sectors were written in the order Taurus, Gemini, Cancer (although we shall consider them in reverse order). These were produced over several hours and represent normal observations concentrating on planet and cloud positions. Aries remains enigmatic. Pisces and the Path sectors show the trajectory of the asteroid as it crossed the sky. The Plume and North sectors record post impact phenomena but also complete the circle across the entire sky.

3

THE CANCER SECTOR

The region of sky covered in this sector (Figure 3) is around the constellation of Cancer following the logic of other sectors that follow constellations of the zodiac, which in general seems to be quite close to the modern zodiac. However none of the stars of this constellation or its name are mentioned, presumably because it is too close to sunrise for the stars in this rather faint constellation to be seen. The discussion is on the imminent sunrise, cloud and planet positions and timing. The timing and dating is presumably recorded in this sector as time was referenced to the motion of the Sun - in particular sunset.

As Figure 3 shows the text in this sector has three orientations. The normal and sideways text will be discussed first. The upside-down text will be considered separately as it is proposed that this was added after the sector had been completed while the observer was working on the Pisces and Path sectors.

The lower boundary has the sign ŠAR$_2$ (B396) "horizon" written three times, making the orientation of this sector unambiguous. This is the only sector where the horizon as been marked in this way.

A region of the sky to the top left has the sign AN (B13) "sky" written circumferentially four times, believed to indicate a clear sky and adjacent to these the sign LUGAL (B151) "king" written four times in a rectangular array to indicate four bright and important objects, probably the main stars of Gemini, the subject of the adjacent sector. There is frequent cross sector referencing on the tablet. The term LUGAL is frequently used in cuneiform with a determinative prefix but when the context is known the determinative prefix is often dropped. Here the prefix would almost certainly have been MUL making "king star".

The upper boundary of the sector is marked by five AN (B13) "sky" signs, again indicating a clear (and therefore starry) region of

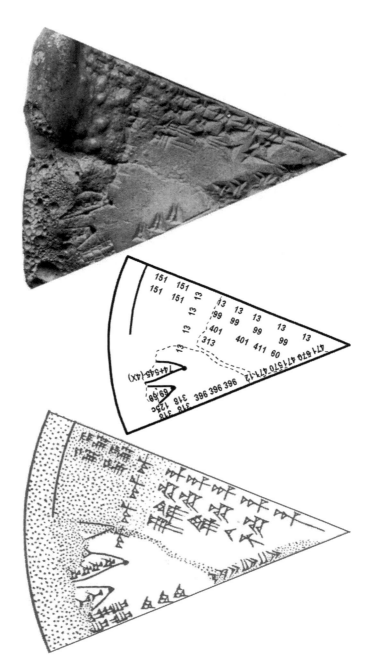

Figure 3: The Cancer Sector

sky. Then beneath this the sign EN (B99) appears four times in a line; it is speculated this may indicate four planets are visible.

The use of EN to mean "planet" is not usual, but associations of the planets, including the Sun and Moon, with a name containing EN is known, or inferable. Thus, Brown [16] shows that the general name EN.TE.NE(NA).BAR.GUZ(ÞUM) has been applied to Mercury, Venus and probably Jupiter and Saturn. This can be translated as "EN encounters bright shaggy smoke", clearly a reference to entering the region of the Milky Way. NA is taken to refer to "fine smoke" such as produced by incense, a meaning that is discussed further in Chapter 5. In the MUL APIN the constellation Crux is also called ENTENABARÞUM and rises Helically as the Sun enters the Milky Way. The original Sumerian name for the Moon was EN.ZU. Mars is the only planet not recorded to have an EN association, but it is a reasonable assumption that it had one based on the above discussion. The EN appellation is encountered again in the Pisces sector in connection with the asteroid.

EN has a known interpretation as "diviner" and the use of planets in the Assyrian period as objects of divination have been explored by Brown [16]. It is therefore plausible to associate EN with planets in the current context, an assumption that makes sense.

An alternative meaning of the EN signs might relate to the root of the word E = wall and N = high, thus EN is a high wall. The original Sumerian pictogram (Figure 4) confirms this as a defensive wall and presumably EN meaning "Lord" originally referred to men who could be trusted with the defence of the city's walls (the title "Marquess" in English has a similar derivation). In this context the observer might be indicating the cloud wall of the warm front that is coming in from the south-east. The orientation of the text relative to the horizon fits this interpretation better in this sector. However it should be noted that EN translated as a cloud wall creates a problem in the Pisces sector.

Figure 4: Early Pictogram for EN (B99) [13] *Suggestive of a high defensive wall.*

Below this line of EN's there are two lines of signs;

MUR (B401) MUR (B401) ŠUŠ (B411) KUR₂ (B60) // KID (B313)

which translates as "region overcast by a strange (reed?) mat" suggesting heavy mottled cloud cover in this region.

Below this text is a diagram showing two 'dots' (actually circular impressions made in the clay by the stylus end) these are marked with zones leading to the left, each containing upside down text. Although not labelled these must be planets as the stars of Cancer would not be visible 24 minutes before dawn, a conclusion reinforced by the angle made by the two dots relative to the horizon being consistent with the ecliptic. For the preferred date of 3123 BC these planets would be Mercury (lowest) and Jupiter and the match of their positions with the tablet is good as shown in Figure 5

The remaining text in the Cancer section is upside down not only to this sector but to the logic used in the rest of the tablet. This upside down text provides an insight into the sequence in which the tablet's sectors were written. The Cancer (normal text), Gemini,

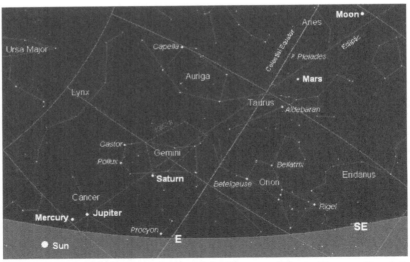

Figure 5: Redshift 5 Simulation of the Region of Sky in Cancer, Gemini, Taurus and Aries for 29th June 3123BC at 1.23 Universal Time
(Note the location of the Sun below the horizon, Mercury and Jupiter above it, Saturn in Gemini, Mars close to Aldebaran and the Moon in Aries. Note also the relative locations of Gemini, Orion, Eridanus, Taurus and Aries. This area of sky accounts for 50% of the sectors shown on K8538.)

Taurus and probably the Aries sectors contain no information about the incoming asteroid and were probably written well before its observation. Thus these sectors show us a normal observation of planets, clouds and other normal aerial and astronomical phenomena and it is speculated that these were probably composed in a leisurely manner as the subject matter presented a good view of itself and precise timing was of little value and probably not recorded.

It would appear that once the Köfels' asteroid was observed, the tablet's author worked rapidly on the Pisces and Path sectors to record it (maybe breaking off the work on the Cancer sector before getting around to recording the planets' names). The measurement of the path and in particular the crossing of the celestial equator are observations that require precise timing and these are noted in the Cancer sector (where the Sun is and hence the place to record timings) without turning the tablet round – hence the cuneiform is upside down. The observer also noted the change in cloud cover over the two planets since making his original observation, again without turning the tablet around.

In the lower left corner the signs are BA_6 (B318) BA_6 (B318) "the rising sun". This is unusual usage since the accepted expression would be $BABBAR_2$ (B381), which Halloran [15] notes to be a reduplicated BAR_6. It is found that BA_6 has been consistently employed instead of BAR_6 throughout on K8538.

The line beneath this is BA_6 (B318) $AŠ_4$ (B125c) "the rising sun 6" indicating that the sun is expected in 24 minutes (6 Sumerian time units of 4 minutes, later called an uš in the Akkadian; the Sumerian is not known). The two symbols in this case are very close and are almost merged into one. This merging of symbols is also found elsewhere on K8538.

The zone connecting the two planets is labelled with BAD (B69) BAD (B69) "clear, clear". The zone connecting the upper object contains 4 signs believed to be BAR + $ŠUŠ_2$ (B74 + B545) "thinning overcast, .. , .. , ..). This interpretation, while it makes sense, is not conclusive due to difficulties interpreting the merged symbols.

The following string of numbers is written next to the lower (horizon) boundary:

20 (B471) 120 (B570) 20 (B471) 2 (B570) 20 (B471) %%%, interpreted as "20 (days), 142 and 20/60th (uš), i.e. 9 hours 29 minutes, from the start of the month" (note: units are not explicit for any numbers on the tablet).

This cannot be proved absolutely because Sumerian numbers are extremely ambiguous. As already noted this is adopted first as a hypothesis (the second context assumption) and then shown to result in an accurate interpretation of the astronomy on the tablet. In particular the configuration of the constellations and the location of sunrise fix the date at the June/July boundary (Julian calendar).

The Mesopotamian lunar month started from the first sunset after a new moon when the crescent moon can be seen. The new day started at sunset. The date is therefore the 21st day into the month but to relate this to a modern date requires identification of which year. The tablet records no year or month, which is consistent with its being intended as a temporary record.

The time of 142.3 uš after sunset is 9 hours 29 minutes, which for 29th June 3123 BC (Julian calendar) would be 1.23 UT. It has been assumed that this time relates to the asteroid crossing the celestial equator as noted in the Path sector. Adding the 6 uš (24 minutes) until dawn to the observation time gives a total night length of 9 hours 53 minutes which, as shown in the next chapter, is consistent with a late June or early July date (Julian calendar) in Sumer.

TIMING ERRORS

4.1 ERROR SOURCES

There are two times on the tablet K8538; these are the time to the event from sunset (9 hours 29 minutes) and the time to sunrise (24 minutes) from which a total night length of 9 hours 53 minutes can be derived. This chapter discusses errors in these times to derive an estimate of accuracy. The accuracy can be assessed in three ways: a direct consideration of the sources of errors, the time indicated by the stars and the night timing compared with the location of the observation.

Table 1 shows a summary of the various sources of errors. Because some of the factors contributing to the overall error are systematic and discontinuous, the errors have been simply added to give a worst case value. While there remains some vagueness about the real errors, it is believed that the analysis supports the conclusion of an error range of −10 minutes to +10 minutes on both the observation time and the day length.

Without any detailed knowledge of the timing instrument a direct estimate of the error in the observer's original reading is not possible. The instrument used was presumably some form of water clock and generally it seems a consensus opinion that water clocks have large errors that are in the order of substantial factions of an hour. However this opinion has no firm basis, and certainly such a clock would have no value for astronomical observations.

A water clock's accuracy primarily depends upon the calibration rather than the technology or workmanship of its construction. The tablet indicates the observers had an instrument that marked out the celestial equator and if this were used for calibrating the clock then time keeping was probably of a high order. The instrument

Table 1:   Estimated Errors in Timings on Tablet K8538

Error Source	Observation		Night length		Notes
	- ve	+ve	-ve	+ve	
Original reading	1	1	2	2	Night length has 2 readings
Definition of sunset	1.25	1.25	1.25	1.25	Sunset takes 150 secs
Point of observation	3.34	1.25	-	-	If not Celestial Equator
Misinterpret fraction	1.33	-	1.33	-	20/60 ° 4 min
Missing fraction	-	-	-	3	On sunrise time
Horizon uncertainty	3	3	3	3	+/- 0.5 deg
Sidereal/Solar time	1.56	-	1.62	-	If time is sidereal
TOTAL	11.48	6.5	9.20	9.25	Take as +/- 10 min.

would have been started at sunset and been constantly attended throughout the night, probably including synchronising with the rotation of the celestial sphere. In these circumstances an accuracy of 1 minute over the night would not seem unreasonable – although even this estimate may be doing the observer an injustice. The night length time is the result of two readings, hence the error is doubled.

An alternative to a water clock could be direct timing of the movement of the celestial sphere using the instrument that marked out the celestial equator (which the observer clearly had). This certainly should be capable of time measurement to better than a minute provided that suitable stars are visible.

The quoted error value of 1 minute can be supported from the indirect evidence that can be gleaned from the fraction (1/3) of the standard 4 minute interval on the time since sunset as quoted in the table. This would suggest the observer believed he had a timing accuracy better than 1 minute and if that were not the case observation over many years would have soon highlighted the problem.

All the other errors identified in Table 1 are due to uncertainties in understanding the tablet and the conventions the observer used rather than inherent errors in the original observation.

The Definition of Sunset

It is known that the Mesopotamian astronomers started the "day" at sunset, but in the context of a clock that can measure to better than a minute it is important to know what defines sunset, as the time from it first touching the horizon to the last sight takes 150 seconds. A corresponding problem exists for the definition of sunrise but it is assumed this would match the sunset definition to give an equal day and night times at the equinoxes. It is not known what point defined sunset for the Sumerians (or indeed later Mesopotamian astronomers). The times presented in this paper are from half way through sunset, which minimises the error from this uncertainty but this is almost certainly not what was actually used.

The Point of Observation

It is not clear which point in the path of the object across the sky the time given on the tablet refers to. From the point shown in Pisces, where the object was apparently first observed, to it disappearing due to reaching the earth's shadow (coincidentally within 5 seconds of reaching the horizon) took around 277 seconds (4.6 minutes). The impact was 97 seconds after the object disappeared, which the observer would have noted as the plume rose over the horizon a few minutes after that. The only point noted on the tablet during its path is the crossing of the celestial equator (a point to which the whole Path segment is devoted) and this takes place 202 seconds (3.34 minutes) after the first observation.

Because the construction uses two stars close to the trajectory rather than the trajectory itself we assume the construction shown on the tablet was done after the asteroid was visible. However, using an instrument that maps the celestial equator onto the night sky, the crossing could be very precisely determined. Both the Pisces and the Path segments suggest such an instrument was available to the observer. Because of the prominence given to the crossing and the precision with which it could determined, it is assumed this is the point in the trajectory that the timing refers to.

If this assumption is in error then the range of times could be from first observation until the object disappears, that is –3.34 to +1.25 minutes. It is thought unlikely the timing would refer to observation of the plume and so the additional time has not been included as a likely error.

This error does not affect the total night timing.

Misinterpretation of the Fraction

The event time assumes the time after sunset to be read as 142 + 20 (that is 20/60ths - a third) Units. This last 20 could be a misinterpretation and refer to the time unit rather than a fraction. If the reading of the tablet assumes the fraction were in error then the actual time would be 1.33 minutes earlier than the times presented in this monograph.

Missing Fraction

The section of the tablet dealing with the sunrise is interpreted as saying it is 6 units (24 minutes) away. The cuneiform for 6 is the last before damage to the tablet makes reading the remaining text impossible. It is therefore possible that the next cuneiform was a fraction number (as assumed on the other time). Assuming these fraction terms were simple (e.g. 1/3, 1/2, 3/4) then this means sunrise could be up to 3 minutes later. Obviously this error only affects the quoted day length time.

The Horizon Error

The biggest potential source of error is that the precise location of the horizons used to determine sunset and sunrise is not known. Given the geography of the Mesopotamian region it cannot be the sea horizon that has been used to determine the sunset times used in this paper. While the region is flat, there is likely to be some error due to the difference between the real horizon and the sea level assumption. This is potentially very significant as every degree difference produces an error of 4 minutes.

Since the observation was clearly made at a special observatory (the tablet indicates a location away from the city it was associated with), the viewing is likely to be good. This is confirmed by the simultaneous observation of Altair and the planets below Gemini indicating less than 10 degrees total reduction from a 180 horizon. However consideration of the following points:

- the horizon is used as a measuring reference;
- the site must have been chosen and prepared for observations;
- the geography of Mesopotamia allows for near perfect horizons;
- the tablet hints Sumerians were aware of sea horizons;

would lead to the conclusion the horizon would be much better than that. From these arguments the timing errors assumed here are guessed at 3 minutes corresponding horizon uncertainty of 0.5 degrees.

Sidereal/Solar Time

This work has assumed solar time as it is time used by the software employed in the analysis and is a reasonable assumption as most cultures do use solar time.

However the tablet is the work of astronomers who may prefer the use of sidereal time. Given a water clock of the accuracy suggested above, the difference between the two would have been noticed by the astronomers making the tablet. Some Assyrian and Babylonian eclipse reports are times relative to the meridian passage of stars [17], and the clear implication of an instrument marking out the celestial equator (which would be useful for calibrating water clocks) both suggest use of sidereal time might be a possibility.

If this were the case it would produce an error of around one and half minutes over the 10 hour night. The errors shown in the table are more precisely determined for the time since sunset.

4.2 SKY MATCHING

A separate timing of the sky in the Planisphere can be obtained through consideration of the total sky that is shown. Because virtually the whole sky hemisphere is described, it can established that this sky can have only been seen for a rather short period of time during each summer night.

Considering the preferred date of 29th June 3123 BC and taking the nominal time to the event as 9 hours 29 minutes (which corresponds to 1:23 UT). The planets in Cancer have risen 20 minutes earlier but could not have been clearly seen at that time. To be seen in the way indicated in the Cancer sector these planets the time must be at least 10 minutes after they have risen (about 1:13 UT). In the other direction the sunrise is 23 minutes away (1:26 UT) and even at the nominal time the dawn glow would already be hampering the viewing. To see Cancer and the stars Enif and Altair in the Path Sector, the observation would have to be no more than about 10 minutes either side of the nominal time with a slight preference for the earlier time rather than the later.

Very similar arguments can made for all the possible dates. Thus consideration of the timing constraints to see the stars and planets shown suggests a window of around 20 minutes and supports the error analysis accuracy of +/- 10 minutes.

4.3 LATITUDE DETERMINATION

The length of night as indicated by the tablet is 9 hours 54 minutes (i.e. 594 minutes). For a given date this can be used to provide an estimate of the latitude of the observation.

Figure 6 shows the length of night against latitude for 29th June 3123 BC as measured by sun half way through sunset to half way through sunrise. The nominal time indicated by the tablet and the 10 minutes error, which includes an allowance for different definitions of sunset, are shown and it can be seen that it corresponds to latitudes between 30° and 34°.

When compared with the various Mesopotamian sites listed in Table 2 it can be seen that the observation is made in the region of Sumer or southern Akkad with a preference for centres such as Kish or Nippur – the analysis in this work has assumed Kish is the location but the conclusions would be the same for anywhere in Sumer.

However we can rule out the observation being made in the more northerly centres of the later Mesopotamian empires such as Ashur and especially Nineveh, where the tablet was found.

That the timing suggests the observation was made from Sumer is not a surprising result and it is argued that the important conclusion is that this result gives further confidence in the interpretation of the timing and also in the 10 minutes error estimate.

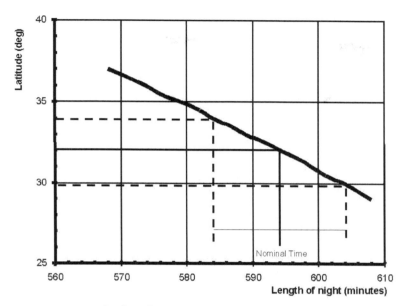

Figure 6:   Length of Night Against Latitude for 29th June 3123 BC

Table 2:   Latitudes of Mesopotamian Sites (from [18])

Nineveh	36°24'
Ashur	35°29'
Kish	32°33'
Nippur	32°10'
Uruk	31°18'
Ur	30°56'
Eridu	30°50'

4.4 TIMING ERROR CONCLUSIONS

It is believed that the original observation was timed to a high degree of accuracy – to within around one minute - but the uncertainties in the conventions used by the observer bring the accuracy that the time can be established to within +/- 10 minutes. By normal standards of dating past events this is an extraordinary degree of accuracy but it is supported in three ways:

i) the analysis of known error sources;

ii) the short time that the stars shown can be seen together; and

iii) the consistency of night length with an observation from
     Sumerian latitudes.
When these arguments are considered together it is believed that
the error estimate has a solid basis.

5

THE GEMINI SECTOR

The Gemini sector (Figure 7) seems to have been primarily intended to show the position of planets relative to the fixed stars.

Along the top boundary the word AN (B13) "sky" is repeated 10 or possibly 11 times indicating a starry clime. This follows the same pattern as, and agrees with, the Cancer sector, which indicated clear sky in this direction. Beneath this is the sequence of text;

MUL (B129a) MAŠ (B74) TAB (B124) BA (B5) NI$_3$ (B597)
AŠ(B1) ŠI (B449) AD (B145)

that is "the constellation of the young twins host the one who has not shied" (or "does not shy"). That the "young twins" is Gemini is not only consistent with the modern zodiac constellation name but is confirmed by interpretations of the MUL APIN text. It should be noted the homophone NI$_2$ (B399) has been assumed to be intended instead of the NI$_3$ written down in order to make sense of the statement.

This text probably refers to a planet, although the early Sumerian names for planets are not known for certain, and AŠ - ŠI - AD does not correspond to the later names that are known, so a positive identification is not possible. However for the preferred date in 3123 BC the planet in Gemini is Saturn. There are two known Sumerian names for Saturn GENNA (a possible borrowing from Akkadian [15]) or SAG.UŠ(which is the term used in the MUL APIN) both of which mean "constant" or "The Steady One". Brown [16] discusses the Akkadian perceived characteristics of the planets and argues that Saturn's various names reflect the planet's slow and deliberate movement. While the term AŠ-ŠI-AD, "not shying" is not the same as "constant", it is a similar concept and we may be seeing an early version of Saturn's name.

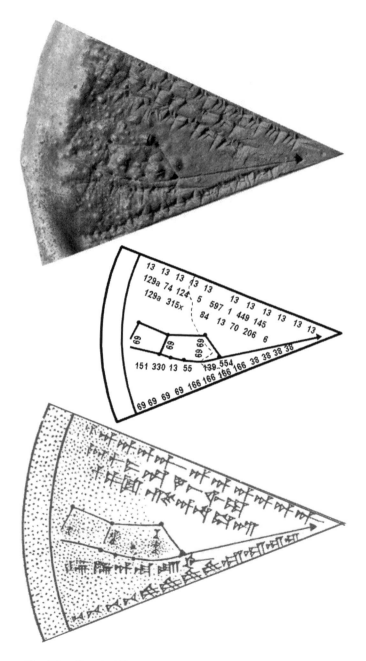

Figure 7: The Gemini Sector

An alternative explanation, which is consistent with the planets not being named, is that the Sumerian astronomer's familiarity with the sky made it unnecessary, is that this text is indicating that there is a planet in Gemini that has not retrograded its motion. For the preferred date of 3123 BC Mercury was also in Gemini a few weeks earlier performing retrograde motion, but was now in Cancer. A month after the proposed date Saturn also retrograded, moving temporarily out of Gemini before returning.

The authors' opinion is that it is unlikely that the normal practice in recording the night sky would be to not name planets when they are drawn and this means they have a strong preference for the first explanation.

Beneath this there is another sequence of signs that reads:

MUL (B129a) SIPA (B315x) ZI (B84) AN (B13) NA (B70)
GUB(B206) ZU (B6)

"the constellation of the faithful shepherd of the Milky Way stands in knowledge", indicating the 'province' over which Orion dominates. As with Gemini the identification of MUL SIPA ZI AN-NA with Orion is confirmed by the MUL APIN text [6].

The purpose of this line is to link Gemini to the triangle representing a planet and define their separation distance. The line does not scale with the constellation drawing. We presume, as the planet is crammed into the pointed end of the triangle-shaped sector, that there was not enough room to draw the stars in, but the words still enable the constellation of Orion to be used as a description of the separation of Gemini and the depicted object. And indeed for Mars on the 29th June 3123 BC it is a good indicative guide.

Turning to the term AN-NA, it is generally interpreted as meaning heaven (i.e. a synonym for AN) [15] and some compound words, such as GI_6-AN-NA meaning "at night" ('black' + 'heaven' + locative) [15], give support to this interpretation. However AN-NA is composed of two symbols (B13 + B70) and it is argued this should be read as AN (heaven) and NA ("incense" or "fine smoke"; [13 and 15]) and refers specifically to the Milky Way rather than heaven as a general concept. NA (B70) in the sense of fine cloud (from the fine smoke produced by burning incense) is also used in both the Pisces sector and the Path sector to describe clouds. The constellations with AN-NA as part of their name all lie either in the Milky Way or boarder it, supporting this identification. Thus Orion would be "the faithful shepherd of the Milky Way", not heaven, and GU$_4$AN.NA would be "the bull of the Milky Way", not heaven etc.

The centre of the sector has a sketch of the Gemini constellation. The stars that form the constellation drawing, in their actual 3123 BC positions, are shown in Figure 8. The clear use of Gemini 1 (a relatively dim star) is a little surprising; otherwise the outline makes sense even if it is not quite as modern astronomers see the constellation. The lower part of the drawing seems less accurate than the upper part, particularly the position of the star Mekbuda and the apparent use of 36 Gemini (close to the limit of visual magnitude) is surprising. However this region is damaged so the extent of the inaccuracy cannot be reliably established. Where the tablet is not too damaged a comparison of the line length and angles on the drawing and the real night sky are shown indicating a good quality hand sketch even after being copied.

Both the Gemini and North sectors contain what to modern eyes looks like "conventional" constellation drawings with the stars represented as dots and lines between them making the overall shape. These shapes seem to be standardised as elsewhere on the tablet the observer seems quite happy to draw these standard shapes even if he cannot actually observe the stars in them. When this is done the convention seems to be that the star dot is left out. This happens in the North sector where dimmer stars are not shown and in the Pisces sector where an entire constellation is drawn without star dots.

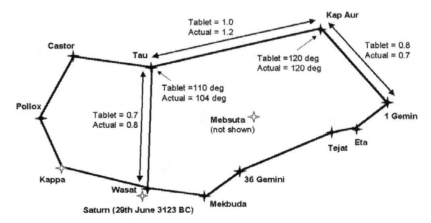

Figure 8: Gemini Drawing Compared to Actual Star Positions

In the Gemini sector Gemini 1 (with magnitude 4.18) is certainly shown as visible and possibly Gemini 36 (below magnitude 5) is also shown. If these very dim stars can be seen it indicates very good seeing conditions from which it can be deduced that this was observed well before the near dawn sky shown in the Cancer sector.

In addition to dots, a few celestial objects are marked as triangles. To modern eyes these triangular impressions look like arrowheads implying directionality to the line. However there is nothing to indicate this is what is meant and most past studies of Tablet K8538 from King [3] on have agreed that the 'triangles' most likely represent bright objects. However now that we have a more complete identification this interpretation is not strongly supported. Table 3 lists the use of the triangle in the tablet's pictures and, if Gemini 1 is neglected, the objects depicted are all bright but much brighter objects, both stars and planets, are shown as dots.

In both the Gemini and North sectors, where triangles are used in constellation drawings they are on lines that do not scale with the rest of the drawing and where text is used to indicate the distance. So this might be the meaning, but it is not consistent with the use in the path sector, which scales very precisely. Nor is it consistent with the use of solid triangles within a cartouche in Taurus. Another possibility that has been considered is that the triangles indicate objects (or lines attached to them) that have text associated with them and this would be consistent with use in the Gemini and North sectors and the trajectory line in the Path sector but not with the other two lines in that diagram that also have associated text but end in dots (but not star dots). This also is not consistent with the use of solid triangles within a cartouche in Taurus.

The favoured explanation, assuming the Gemini 1 triangle is ignored as damage, is that the triangles are used to depict celestial objects that are not part of the constellation depicted or otherwise

Table 3: Use of Triangle to Denote Objects

Sector	Object	Magnitude	Notes
Gemini	Gemini 1	4.18	"Triangle" might be damage
	Taurus Planet	0.51	Mag. for Mars in 3123 BC
Path	Enif	2.37	
	Altair	0.76	
North	Capella	0.08	Tentative identification

shown alone. This is consistent with all uses on the tablet but does raise the question as to why the planets in Cancer are not then depicted as triangles. Thus, while it is certain the triangles are used to denote celestial objects in the diagrams that can be interpreted, the convention for their use instead of dots has eluded convincing interpretation.

The drawing of Gemini was presumably to show the location of the planet but this information is now lost due to the damage, which would be consistent with all the drawings on K8538 recording transient events. The ecliptic runs through the lower part of the constellation and it is possible that either of the two dots, which we have identified as Mekbuda and Gemini 36, could actually be the planet. Both of these identifications have reasons to be queried, as Mekbuda is not located with the accuracy shown elsewhere in the drawing, and Gemini 36 is a very faint star (magnitude 5.26). However both seem to be connected in the constellation line drawing and therefore are presumably fixed stars. It is thought unlikely that a planet would be incorporated as part of a constellation outline.

It must be concluded that no convincing identification of the planet's position can be made. The discussion of how this impacts the determination of the date is contained in Chapter 9

Within the sketch of Gemini there are four regions labelled BAD (B69) "clear" indicating lack of cloud within the region. Since it is obvious that the sky was clear when the original observation was made, this note is likely to be a later addition. It is speculated that these extra words were added at the same time as the upside down notes in the Cancer sector (the original cuneiform B69 is axisymmetric and may also actually have been upside down originally). In which case the notes have been added to indicate Gemini was still free of cloud when the Köfels asteroid was observed.

Under the diagram of Gemini is the sequence:

LUGAL (B151) LU$_2$ (B330) MUL (B13) LA (B55) TA (139)
SAL(B554)

which translates as "king star of abundance through persistence". In the MUL APIN the term LUGAL LAL is used as a name for Pollux the brightest star in Gemini. However the phrase "star of abundance through persistence" is applied to several stars and it is proposed here that it is a term applied to the last star to fade in the morning. As such, at different times of year different stars would have had the accolade of AN LA TA SAL. Presumably one of the astronomer's jobs would be to note the last star to fade for that night and this shows he has done that, in fact he has done this well before other stars have disappeared, which suggests a somewhat casual attitude to this particular task.

An alternative, but unlikely, possibility is that the location of this text is next to the planet in Gemini and that it refers to the planet and not Pollux.

It should be noted that the translation of B13 as MUL is not normal and required explanation. B13 is known to represent either AN "heaven" (maybe from A "Water"+ N "High" [15]) or DINGAR "deity". In early Sumerian, pictogram B13 was denoted by as a single star (Figure 9). However this pictogram of a single star was also one of several ways to denote MUL (B129) [13]. MUL (maybe from MI, "night" + UL, "ornament" [15]) can mean either a single star or a constellation (many stars) and it reasonable to assume that the single star pictogram originally meant MUL (singular) and the multi-star pictogram originally meant MUL (pural) i.e. constellation. Later cuneiform used the evolved multi-star version for both MUL (singular) and MUL (pural) and it may not have occurred to the Assyrian copier that a single star in this context should be treated as MUL.

This is consistent with the fact that in tablet K8538 the multi-star logogram (B129a) always means MUL as constellation. In the sectors previously discussed B13 has been used in multiples to indicate AN as "heaven". Indeed this way of using it by Sumerian astronomers may be how the star symbol acquired the "heaven" meaning. However here, and in the Pisces and Path sectors, B13 (used on its own) is clearly intended to indicate a single star: that is MUL (single). If this interpretation is correct, it implies the original tablet was written in the early Sumerian Pictogram form of cuneiform.

	B13	B129a
EARLY SUMERIAN	✳ AN DINGAR MUL (Single)	✳✳✳ MUL (Pural = Constellation)
LATE ASSYRIAN	►►┬ AN DINGAR	►►┬►►┬ MUL (Single and Pural)

Figure 9: Evolution of the MUL Logogram [13]

Along the lower boundary of the sector some ground features are named although this is not thought to represent the horizon plane because the outline of Gemini would not then have the correct orientation to the ground. Also it is not labelled as the horizon, whereas in Cancer the lower line is so labelled. The lower boundary on the far right has the symbol URU (B38) "city" repeated four times, and next to that moving to the left is the word KASKAL (B166) "road leading" inserted four times and finally to the left of that BAD (B69) "open country" inserted four times. It seems that there is open country with a road leading to a conurbation to the east of the observer, which is visible in the early morning. This is a strange thing to record and it can be speculated that this might be essentially doodling while waiting for the Cancer planets to rise. At four o'clock in the morning after a long shift the location of an inviting warm bed in the city (and the associated attractions) might be preying on the observer's mind. If this is the case it is a further indication of the original being written in a pictogram written language (early Sumerian) rather than an abstract written language (later cuneiform).

Whatever the reason for its inclusion, this line gives an interesting insight into the location of the observatory. It shows that it was outside, and to the west, of the associated city, presumably by several kilometres.

6

THE TAURUS AND ARIES SECTORS

6.1 TAURUS SECTOR

The Taurus sector (Figure 10) is badly damaged with about half of the area missing. Its full purpose is therefore more speculative, but most likely it is discussing the planet the Gemini sector shows as being there.

The lower boundary of the sector locates Orion with the sequence:

MUL (B129a) SIPA (B315x) ZI (B84) AN (B13) NA (B70)

"the constellation of the faithful shepherd of the milky way" and as noted in the Gemini sector above, this is known to be Orion from other texts such as the MUL APIN.

Four lines of text are oriented parallel to the upper boundary of the sector and lines 3 and 4 read:

AN (B13) NI (B231) NI (B231) // AB(B128) ŠITA$_3$ (B83)

"the sea channel of heaven" and is identified as the constellation now known as Eridanus, located next to Orion in the position indicated. As an aside, once this epithet is applied, it is interesting to consider the resemblance of northern Eridanus to the Persian Gulf.

The sequence of signs in lines 1 and 2 then reads:

%%% SUKKAL (B321) AN (B13) NA (B70) // AN (B13)
SAKKAR (B212) SILA (B12)

which translates as "envoy of the Milky Way which divides the sky dust".

The MUL APIN also refers to 'PAPSUKKAL' (first envoy) of Anu and Ishtar which is nearly a literal translation of the above so that the %%% sign may be PAP "Lead". From its location within the sector it is apparently referring to the Pleiades as the first envoy (which rises just ahead) of the Milky Way.

37

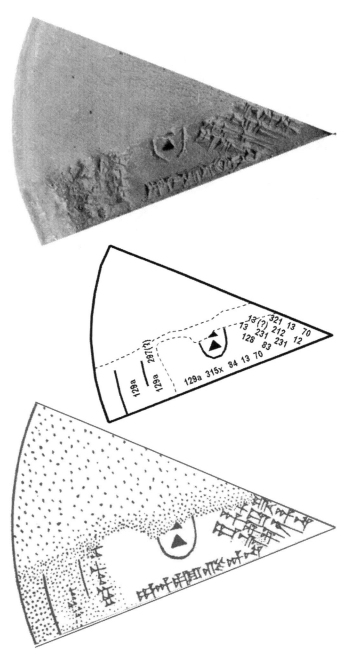

Figure 10: The Taurus Sector

Significantly, the region of Libra in the MUL APIN text had the constellation name ZI.BA.AN.NA that translates as "true division of the Milky Way". This location is 180° along the Celestial Equator from the Pleiades and on the edge of the Milky Way. Thus the Pleiades are the '(first) envoy' and there is an antipodal location called the true division. The question of the role of the Pleiades as a zero point for the celestial equator is discussed in the Path sector (Chapter 8).

Along the periphery of this sector can be identified MUL B(129a) %%%. "constellation ??". The very damaged symbol cannot be read except for the leading cuneiform strokes. These are not inconsistent with it being GU_4 (B297), part of the constellation of Taurus, which would be expected. Other constellations mentioned in the MUL APIN text in this region equating to Auriga, Perseus and 'the crook staff' are not good contenders for what can be read of this sign.

An enclosure dominates this sector with at least two objects inside. This could be the object shown in the Gemini sector. It probably encloses Aldebaran and a planet, a hypothesis supported by modelling the proposed dates using Redshift. For the 3123 BC date the planet is Mars and both objects would be very similar in colour and magnitude. There is also an interesting near alignment of Aldebaran, Mars, and the Pleiades, which may have been of interest to the observer. The damage in this sector unfortunately erases any further information so the complete purpose of the observation is lost.

6.2 ARIES SECTOR

This sector (Figure 11) is very badly damaged but appears to have contained ten identical, radially written sequences:

UDU (B537) ŠUB (B68) DI (B457) "the sheep casts judgement".

However DI is polyphonic with SI_8 and it seems more probable that the homophone SI (B112) "light" was intended instead of SI_8. The text then reads "the sheep casts light" which would be a good description of moonlit clouds in Aries where the moon would be in its final quarter based on the stated date placing it during the 21st day of the month. The Sumerian month began at sunset on the day that the moon first passed new and appeared as a slim crescent.

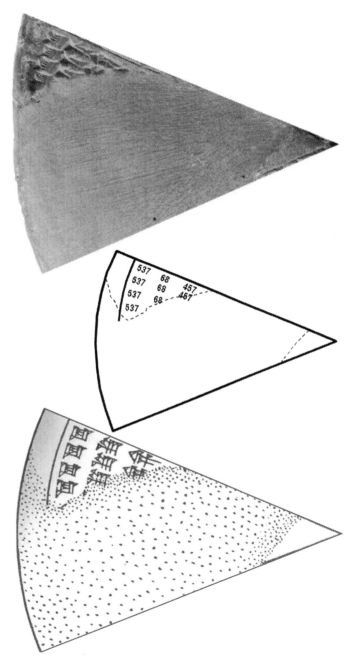

Figure 11: The Aries Sector

The location of the moon in Aries is supported by the astronomical modelling and references to clouds elsewhere shows that there was cloud cover in this part of the sky.

The use of Sheep (a meaning close to, but not, "ram", which would be UDU-NITAP) for Aries would be consistent with MAŠ TAB BA (the young twins) and GU_4 AN-NA (Bull of the Milky Way); that is Aries is another zodiac constellation that has essentially the same name today as was used in Sumerian times. It seems plausible that the constellation of Aries was then recognised as a ram since the Babylonian calendar had month 1 bar.zag.ga [19] "the sacrifice of righteousness". Ovine sacrifice was an ancient custom in the Middle East as attested by the Bible where this same phrase is used for such a sacrifice.

However to counter this, it should be noted that there is no constellation of "the sheep" or close equivalent mentioned in the MUL APIN text. The normal modern identification from this text is that Aries is the "Hired Man" [6] being the constellation described in the MUL APIN as being between the Pleiades and Annunitu. As a further complication, the Pisces sector indicates that MUL Annunitu is the next constellation along the celestial equator where Aries (or may be Triangulum) ought be. This makes identification of constellations difficult in this region.

A further problem is that every other place on K8538 where a constellation is named, it is preceded by MUL (B129a). To be consistent the translation should read MUL UDU if it does mean constellation of the Sheep.

Given the use of an uncertain constellation identification and a polyphone followed by a homophone mistake, the interpretation of this section is acknowledged to be weak. With so much of the sector missing it is difficult to further this translation even with the near certain knowledge it is dealing with the location of the Moon in Aries covered by cloud.

7

THE PISCES SECTOR

The Pisces sector (Figure 12) is the first of the crucial sectors of K8538 relating to the observation of the Köfels asteroid.

The drawing shows the distinctive shape of Pisces, recognisable as an almost modern rendering of the western half of the constellation as Figure 13 shows. It shows none of the stars as dots (as did the Gemini drawing) and part of it is 'shaded' for lack of a better description. It is probable that this indicates that the constellation was only vaguely, if at all, discernible as a result of the sky brightening before dawn, Pisces having no bright stars. In the North sector there is also a clear indication the observer is prepared to indicate stars he cannot see by intersecting lines without dots. The shading is consistent with indicating the cloud cover as defined elsewhere.

Along the upper boundary of the Pisces sector are four signs, NA (B70) "smoke", a reference to light smoke-like cloud. Below that is the sequence A (B579) NA (B70) "water smoke", possibly heavier steam-like clouds, and finally A (B579) NA(B70) NU (B75) "no water smoke" i.e. no clouds.

Two peripheral sequences of text are lost into the damage, but the outermost one shows:

MUL (B129a) %%% that is "constellation ??"

and the second:

MUL (B129a) A (B579) NU (B75) NI(B231) %%%.

This is probably the constellation recorded in the MUL APIN text as Annunitu although as discussed above in the Aries sector a precise identification is not possible.

The line connecting Pisces to the closed triangle has the sequence;

MUL (B129a) AŠ (B1) KAR$_2$ (B105)

which translates as "constellations the one encircles"

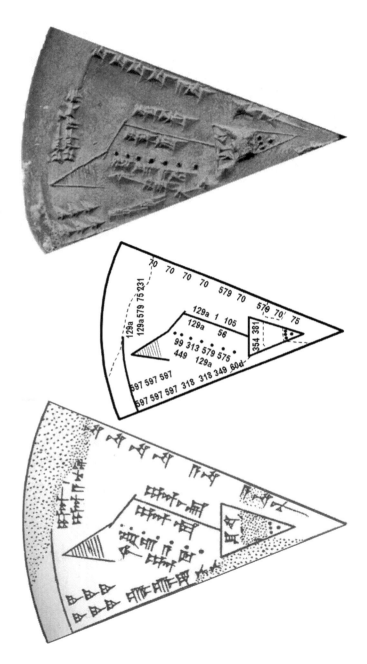

Figure 12: The Pisces Sector

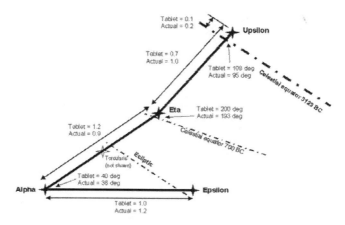

Figure 13: Pisces Drawing Compared to Actual Star Positions for 3123 BC

This can only be the Celestial Equator and, assuming the four 'dots' arranged in a triangle in the closed triangle represent Aquila in the immediate locality of Altair, it would seem the closed triangle is an insert of the next (Path) sector that covers the same area of the sky encompassing part of modern Pegasus and Aquila.

Beneath the Celestial Equator is the sequence MUL (B129a) APIN (B56) "the constellation of the plough", which is the starting point for the MUL APIN text's description of the sky. Past attempts to identify what stars form this constellation [6,7], while agreeing that it is close to Pisces, do not match Figure 12. However there is a strong indication this identification is correct by considering the early Sumerian Pictogram for APIN (B56) shown in Figure 14. This shows the similarity with the picture and the star pattern and is presumably why the constellation was so named.

All that has been said of the Pisces sector so far is known astronomy, but from this point on the tablet starts to record the extraordinary. The next line explains why the constellation was drawn

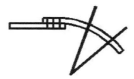

Figure 14: Early Sumerian Logogram for Apin (Plough) [13]

and hence the main purpose of this sector, which is the description of a large bright object crossing the heavens.

Below, and parallel to, the Celestial Equator is a sequence of seven indentations indicating the path of an object, possibly through the broken and thin cloud described against the upper boundary. Beneath this pattern of 'dots' is the two line sequence:

{MUL (B13) EN (B99)} LIL$_2$ (B313) A (B579) TEŠ$_2$ (B575) //
IGI (B449) MUL (B129a).

A conventional translation of this line would be DINGAR ENLIL A TEŠ IGI MUL "the god Enlil passed this way before the stars (or constellation)". However this translation is not consistent with the belief that that the tablet only contains objective observations (the fourth context assumption), as it implies that the observer made an identification with a specific god as he observed it, which is thought unlikely. However this more conventional translation is not inconsistent with the identification of the line of dots as being the asteroid's path.

An alternative meaning can be established by considering in more depth the AN EN logograms. They have been run together indicating they should be read together. Considering the B13 symbol first, there are two alternative ways of reading this symbol. It has already been argued that the B13 symbol can mean MUL "star" (singular) and it may be that this is what was intended in this case. Turning to EN, the use of EN (B99) meaning "planet" was discussed under the Cancer sector. As with some other signs in places on K8538 the AN and EN have been run together indicating a moving star-like object which the observer is having trouble identifying.

The sequence then reads "the star/planet of the wind approached each of them in front of the constellations". Here the " …each of them …" are the positions indicated by the dots on the drawing.

The word LIL$_2$ (B313) is usually taken to mean "wind". However this should not be taken literally since the indications are that the wind or air was perceived as the stuff, or region, which filled the space between the Earth (KI) and heaven (AN). It seems likely that the word is used here to represent an object that has left the heavens and entered the domain of the LIL. Obviously this was a very unusual and special event.

The line of seven dots implies the asteroid was first seen when it was in Pisces. At this point it is estimated it would have had a magnitude of –4.9, which is extremely bright and on a clear night the object should have been seen much earlier than this. The radian point was in the constellation Eridanus and the path then travelled through Aries. The fact it does not appear to have been seen before Pisces is further evidence that there was cloud cover in Aries.

The line of dots does not match well with the trajectory shown in the Path sector. The object was actually some 15 degrees lower in altitude and appeared just below the constellation not within it. However given this was the first sighting (without warning) and that the stars themselves could not be seen, this is not a surprising error and in the circumstances is a good observation. Because the first sighting is thought unlikely to be good compared with the later observation in the Path sector, this has not been used to aid determination of the trajectory.

The lower boundary of the sector is marked by a sequence of text that appears to describe the body. At the extreme left is a 2 x 3 array of the symbol NINDA (B597) one of the meanings of which is 'arc-minute', and may indicate a record of the size of the object as 2 x 3 minutes of arc, an attempt to illustrate the visual dimensions of the body which was not a point light source throughout. The size is consistent with a body a little over 1 kilometre in diameter as supported by the Köfels site evidence. The use of an array in this way is yet another indication that the original tablet may have been more pictorial than written text.

The next part of the sequence:

BA_6 (B318) BA_6 (B318) BUR (B349) DIM_4 (B60d)
translates as "bright white stone bowl coming towards" an attempt to describe the visible appearance, shape and motion of the object. Note that (B60.B60=B60d) and are read together as a single word DIM_4.

Simulations of an assumed spherical object under illumination by the Sun on a trajectory consistent with observation from Mesopotamia show that this is a good description of the body, which would have displayed a small range of phases as it passed by, reminiscent of a simple white hemispherical bowl of the period. This description is consistent with the visual size 3 minutes of arc; that is, an overall shape can be made out but surface features would not be able to be discerned.

The celestial equator line then goes into a triangle. The region in this inset triangle is enlarged in the adjacent Path sector. The object became very bright (magnitude − 6.6) as it moved into this region near Aquila where the sequence ŠU (B354) UD (B381) "pours light" is visible. Other information in the inset cannot be read due to damage.

Taken as a whole the Pisces sector gives an overview observation of the object, but without the detail needed to make a precise determination of the trajectory (that detail is in the Path sector). It does, with both linguistic clues and the indicated position of the celestial equator, complete the range of pointers to the date of the tablet's observation.

8

THE PATH SECTOR

A ccepting the conclusion that the closed triangle in the Pisces Sector is a "breakout box" for the Path sector (Figure 15), it shows sky in eastern Pegasus and Aquila. It follows the two bright objects shown as triangles are most likely to be Enif in Pegasus and Altair, which are the two brightest stars in that region.

A line through the bright stars is taken to be the path of the Köfels' asteroid, the annotation above the line being:

AN (B13) BANDA$_3$ (B144) SI$_2$ (B84).

It is suggested that the homophone SI$_3$ should replace SI$_2$ and that B13 here means MUL (single) "star" (as already discussed in Chapter 5). This then translates as "star vigorously swept along" and at an intermediate position between the bright stars is the statement AN (B13) GAL (B343) "large star", taken to indicate the position of its maximum apparent size and brightness. This corresponds to the point of closest approach as indicated by the trajectory modelling discussed in Chapter 10.

A further point to note is that the Enif triangle is not centred on the trajectory line. This may be an indication that a "near miss" is being indicated rather than a direct superimposition of the object and the star. The trajectory modelling described in Chapter 10 suggests that miss distances were within half a degree. A modern observation would be concerned with the value of the miss distance to establish the trajectory but to the Sumerian observer it would be the association of the object and star that is important and he is not likely to be concerned with quantitative separation distances as these would have no value to him.

The trajectory line is shown crossed by two other lines almost coincident where they cross. One of these lines is annotated:

GA AN (B350) GUDIBIR$_2$ (KUR (B366) + KUR (B366))

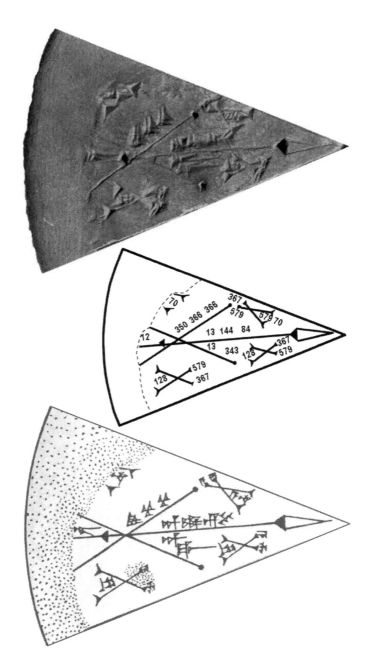

Figure 15: The Path Sector

Which translates as "queen of foreign lands". This line is parallel
to the horizon and appears to intersect the Celestial Equator 90°
west of the Pleiades. The Helical risings and settings in the MUL
APIN text identify a celestial object that lies close to 90° east of the
Pleiades as "the constellation of the King". And, as already noted,
there is ZI.BA.AN.NA "true division of the Milky Way" at 180°
along the Celestial Equator from the Pleiades.

It is proposed that the Pleiades cluster was used by the
Sumerians as the starting reference for angular measurements along
the Celestial Equator and not the Vernal Equinox as used today.
Their name in the MUL APIN text, MUL.MUL, (the "constellation
of constellations") is suggestive of some special prominence. Thus
when the Pleiades are at their Zenith the 'King' point is on the
eastern horizon and the 'Queen' point is on the western horizon,
setting over 'foreign lands'.

The remaining line in the diagram is then readily identified
as the Celestial Equator from astronomy simulation, as shown in
Figure 16. An accurate comparison of the depiction in the Path
sector with the real sky is shown in Figure 17. A small triangle is in
fact formed by the intersection of the three lines, as shown in the

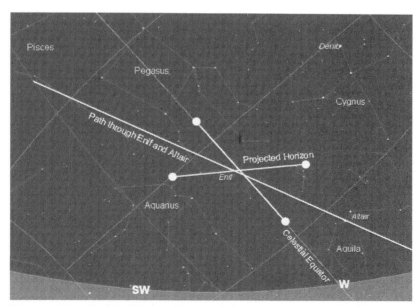

Figure 16: Redshift5 Simulation of the Path Sector with the K8538
Coordinates Superimposed.

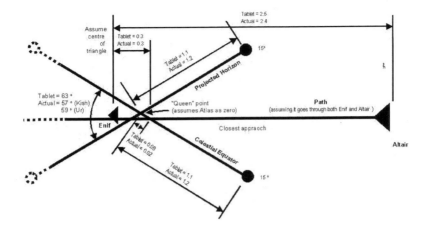

Figure 17: Comparison of Path Sector Drawing and Real Sky.

diagram on the tablet. This small triangle strongly suggests that the star Atlas in the Pleiades was the precise zero point and this is the assumption in Figure 17.

Figures 16 & 17 have also assumed the path line shown was intended to pass through Enif and Altair (how the observer would remember it) rather than the actual trajectory, which varied from this by around 0.5 degrees. The "Queen" horizon line has been shown as constant with azimuth as the celestial equator rotates rather than a projection of the local horizon.

At the extreme left of the trajectory line the text %%% SILA (B12) "divide" indicates that the path approximately bifurcates the angle between the Celestial Equator and the horizontal.

The end point of the Celestial Equator and the horizontal were probably marked by indentations although only two are now visible, the other two being lost in the damaged region. Each of these line end points is annotated with 'crosses' with associated cuneiform symbols as shown in Figure 18.

They state the angular displacement from the reference line. A (B579) "water" is the reference (i.e. level surface) and ŠE (B367) is 1/6th of a right angle to that level surface, i.e. 15°. The angles that are above the reference are labelled as A (B579) NA (B70) "water smoke" or cloud angles, whilst those below the level surface are labelled AB (B128) "sea" angles. ŠE has normally been associated with absolute units [27], but here seems to represent only a fraction, 1/6th.

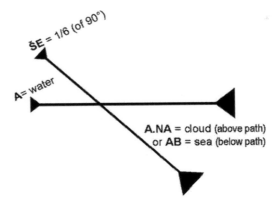

Figure 18:   End Point Cross Sumerian Labels

This suggests that the crosses on the Celestial Equator are there to label the dots at the end of the lines which are 15° above (cloud) and 15° below (sea) the "Queen" point producing a very accurate (although to modern thinking a somewhat strange) scale grid. The scale of the region of sky is thus specified and the identification of Enif and Altair a near certainty as confirmed by the comparison in Figure 17.

The final feature of this sector is the arrowhead triangle that emanates from the Altair solid triangle. This is thought to be another constellation diagram being a part of the modern constellation of Aquila. It is made up of the two stars either side of Altair (Tarazed and Alshain) and, at the point, Delta Aquila. Given the time of night and the low elevation, only Altair would actually be seen so the stars are not drawn in as dots. The same group of stars can be seen in the point of the "breakout triangle" in the Pieces Sector, but here they are depicted as dots without the connecting lines.

9

DATING TABLET K8538

9.1 PLANET SEARCH

The method used to date K8538 was to look for nights with a consistent midsummer date (as determined by the Moon's position) in which the location of the planets match with the position shown in the Cancer and Gemini Sector. The boundary dates for this search were between 3500 BC to 2000 BC. This range was determined by two considerations, the position shown for the celestial equator and the earliest possible use of written language.

As mentioned in the introduction the location of the Celestial Equator in Pisces (Figure 12) is clearly wrong for 700 BC. It is actually consistent with a range between 4000 BC and 2000 BC. It is not possible to be more precise as it is not clear whether the star at the top of the upper line is Tau, Upsilon (as shown in Figure 13), or Phi Pisces. These are all stars of similar magnitude and as the other two are not shown, there no clear reason to guess at one as opposed to others. However, the angle on the bend in the upper line would suggest Upsilon Pisces, which would the best choice for most of the period. The line length between Eta Pisces and the star is more suggestive of Phi Pisces suitable for the late Third Millennium dates, but this forms a straight line with Eta Pisces and Alpha Pisces so the bend shown in the line becomes an issue. The line length between Alpha Pisces and the star is suggestive of Upsilon Pisces and has the bend. This would correspond to late Fourth Millennium or early Third Millennium. These alternatives are illustrated in Figure 19.

Consideration of the position of the Celestial equator in the path sector is also not as discriminatory as might be, though early

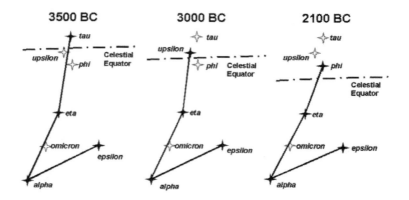

Figure 19: Alternative Interpretations of Pisces Diagram

dates around 4000 BC are ruled out as, at the time, the Celestial Equator actually passes through Enif, which is not compatible with the drawing. However most dates from around 3500 BC could be judged consistent with what is shown. So the search only considered 3500-2000 BC.

Another reason for constraining the search to after 3500 BC is that the Sumerian peoples are not believed to have invented writing until the Uruk period, which Crawford [20] argues is "conventionally dated" starting at 3500 BC (citing Edwards et al [21]) but also suggesting the period might in fact be later. Therefore it seems reasonably certain, whatever later scholarship might uncover, that a cuneiform tablet could not have been produced before 3500 BC simply because that written language had not been invented.

The location of the four planets together with the Lunar date in the Cancer sector is a very discriminatory tool for identifying the date on which this observation was made. Unfortunately the author did not name a single planet (except maybe Saturn), which would have resulted in a single unique date. Without clear labels, however, all the bright objects indicated cannot be unambiguously determined as planets and the object linked to Gemini in Taurus could conceivably (but improbably) be indicating the Sun's Longitude of the Ascending Node, the spring equinox being the start of the Sumerian year.

Table 4 Planetary Matches with Tablet

YEAR	Date	Cancer	Gemini	Taurus	Notes
Good 4 Planet Matches					
2222	29 June	Mercury Jupiter	Venus	Mars	Good match on positions and day lengths
2269	6 July	Mercury, Jupiter	Saturn	Mars	Jupiter and Mercury are rather close together not quite matching tablet. OK time match
3123	29 June	Mercury Jupiter	Saturn	Mars	Good match on positions and day lengths
OK 4 Planet Matches					
2361	4 July	Mercury Venus	Mars	Saturn	Mars not in a good position compared with Gemini sector. Good time match
2625	11 July	Mercury, Jupiter	Venus	Saturn	Venus not in a good position compared with Gemini sector. Timing possible but beginning to push assumed accuracy
Poor 4 Planet Matches					
2769	13 July	Mars Saturn	Venus	Jupiter	Mercury is also in Cancer and to get it below horizon is incompatible with times. Overall timing match poor
2829	16 July	Mars, Mercury	Saturn	Jupiter	Mars and Mercury too flat and too close for good Sunrise match. Saturn is not in a good match for Gemini Sector. Overall timing match poor
3196	26 June	Mercury Venus	Mars	Jupiter	Mercury Venus relationship does not match the Cancer sector at all. Jupiter a little too close? Timing good
3480	14 July	Mars Mercury	Jupiter	Saturn	Venus is also in Cancer and to get it below horizon is incompatible with times. Overall timing match poor
Good 3 Planet Matches- None					
OK 3 Planet Matches					
2210	14 July	Mercury Jupiter	Saturn	-	Good planet match. Timing off

2799	15 July	Venus, Mars	Saturn	-	Saturn not quite right. Timing a little off
3087	19 July	Venus Jupiter	Mars	-	Planets OK but not good. Timing not so good
3181	8 July	Saturn, Venus	Mars	-	Planets OK but not good, Timing good
Poor 3 Planet Matches					
2756	18 July	Venus, Jupiter	Mars	-	Mercury in Leo and visible if timings correct. Overall timings a little off
3092	14 July	Mercury Saturn	Venus	-	Venus is not well placed compared to Gemini sector Jupiter in Aquarius Timing not good
3164	1 July	Venus, Mercury	Mars	-	Planets are very badly positioned. Timing good
3418	18 July	Venus, Mars	Saturn	-	Cancer planets not good. Timing not good
3467	19 July		Mercury, Jupiter	Mars	

The method used was to compare the sky on the tablet including the planetary positions with the night sky as shown by Redshift. It involved looking at the 20th day after the last new moon during June/July period (Julian date) as seen at 1.25 UT from Sumer for all years from 3500 BC to 2000 BC. The planetary positions were noted and compared with those shown in Tablet K8538. The requirements for a match were:

- one planet in Gemini,
- two planets in Cancer,
- no planets in Pisces.

The search found eighteen possible dates and these are listed in Table 4. Not all these sky matches are of equal quality and several wider arguments are used to arrive at the preferred date of 29th June 3123 BC.

While is believed that the object in Taurus is a planet, this is not conclusive and so this has not been used as a selection criterion. Hence both three and four planet matches are identified. However where there is a four planets match the position of the Taurus planet is included in the judgement of quality.

A judgement on the quality of match was made on the basis of the following considerations.

- The position of the fourth planet in Taurus (where relevant) is sufficiently in to merit the description of Orion being between the end of Gemini and the Planet. There is also a preference for it to be close to Aldebaran.
- The position of the planet in Gemini should be to the west of Wasat because this is what the tablet seems to indicate as this area is damaged and does not appear where the tablet is undamaged.
- The positions of two planets in Cancer should align roughly along the ecliptic as indicated on the tablet.

9.2 REASONS FOR PREFERRED DATE

Table 1 shows that on the basis of sky matching alone there are three good candidates all of which are four planet matches.

29 June 2222 BC
6 July 2269 BC
29 June 3123 BC

The 6th July 2269 BC has been rejected as it is late in the year and the trajectory match with a Köfels impact site is bad and there are no other viable candidate impact sites. This leaves 29th June 2222 BC and 29th June 3123 BC. This monograph has used 29th June 3123 BC as the preferred date. This has been selected for the reasons outlined below which includes both evidence internal to the tablet and outside evidence.

9.2.1 Tablet Based Evidence

Trajectory Accuracy

The initial selection is solely based on matching the planetary positions and the constellations. The trajectory adds a further constraint given the end point is Köfels (and there are no other candidate sites). Dates in late June and early July (Julian Calendar) give a much better fit than mid June dates. Given it is not possible to establish the accuracy of the observation and therefore this cannot be used as a definitive case for rejection, it is still reasonable to give the closer matches more credibility, especially considering the apparent high quality of the observation. The 3123 BC date provides a better fit to a trajectory that goes through Köfels than 2222 BC.

Position of the Celestial Equator

As discussed above, the star at the top of the APIN (Pisces) constellation is ambiguous. However when accounting for the bend shown in the line between Alpha Pisces and the star there, there is a preference for Upsilon Pisces which corresponds to the late Fourth Millennium date rather than the late Third Millennium date.

Further in the Path Sector, while the position of the celestial equator in relation to Enif could be argued to match the drawing, the best fit (as shown in Figure 17) is for the period at the end of the Fourth Millennium BC.

The Gemini Planet

As discussed in Chapter 5, there are two proposed explanations for the description of the planet in Gemini as AŠ-ŠI-AD (one who has not shied or one that does not shy – as in the sense of balking).

One explanation is that "one that does not shy" is simply a more ancient way of expressing Saturn's later name "constant". This can only apply to the 3123 BC date and not 2222 BC where the planet in Gemini is Venus.

The alternative explanation for the expression is that it should be read as "one who has not yet shied". That is, the observer is looking out for retrograde motion of the planet concerned. In 3123 BC Saturn is a few days away from a period when it appears stationary just below Gemini (balking) and although true retrograde motion is a few months away the observer may be looking out for this apparent halting. In 2222 BC Venus has no retrograde motion around the time the observation would be made. If this is what is meant, then again it suggests the 3123 BC date.

However there is a difficultly with the position of Saturn. As discussed in Chapter 5 it is not possible to determine where the planet in Gemini is due to the extensive damage in the lower half of the diagram, which is where any planet would be located. For the 3123 BC date we would expect Saturn to be shown to the left of Wasat, a region that is very badly damaged and there is no evidence of a dot.

If we take the interpretation that the planet is one of the two dots identified as Mekbuda and Gemini 36; this makes the position of Mekbuda a very bad fit (as opposed to just a bad one) and Saturn is on the wrong side of Wasat and at least 2 degrees out from the position indicated by the Redshift simulation. As noted Redshift planetary positions are thought to be accurate to a quarter degree and so this is a matter of some concern. This region of the tablet has been damaged so it may be that it is being misinterpreted.

The position of Venus in 2222 BC is better if it is interpreted as the dot after Mekbuda, which we have identified as Gemini 36. But it is thought unlikely that a planet would be part of the constellation.

The actual position of the planet in Gemini remains a problem but without a clear undamaged version of this sector it will not be possible to resolve this issue. The position has not therefore been used as a discriminator between the two dates.

Use of Language

During the translation of K8538 we argue that, while the use of language is consistent with the either end of the Fourth Millennium BC as well as the Third Millennium, there are some features that are more suggestive of the early date.

The use of repeated logograms, in particular the AN (B13) Star to represent open (non-cloudy) sky and the possible doodle in the Gemini sector, suggests the use of pictograms symbols being used as pictures as opposed to pure symbols for words which would indicate a Fourth Millennium BC date.

The apparent use of a single star symbol to mean MUL (singular) for "star" is another pointer to the use of the earliest pictogram form of cuneiform and hence a Fourth Millennium BC date.

9.2.2 External Evidence

Sumerian History

3123 BC would represent the move from the Uruk period to the Jemdet Nasr period about which we have very little real history (the periods are defined by archaeological evidence). So although the earliest writing is extant it is not surprising that there is no definitive record. The 2222 BC date is in a recorded period during the reign of the Kings of Agade and it is less likely any contemporary record would not be known to us.

If this event was the source of myth of the god Enlil coming to Earth and taking over the Universe from the god An, then this would constitute further evidence for the earlier date. It is generally agreed [9, 22] that Sumerians understood the change of god to be an event that had occurred in historical times and that in the Uruk period the chief god was An. By the middle of the Third Millennium BC Enlil had become the top god. Given the change of god was clearly well before 2222 BC, then this date can be ruled out for the impact event, if the myth identification is correct.

Köfels Site Dating

The dating of the Köfels site is discussed in Chapter 12 and concludes that the dating techniques which give ages around 9000 years are unreliable so there is no conclusive dating of the site. However, apart from one fission track date of around 3500 years ago, all the dating suggests an older rather than a younger date more in line with 3123 BC than 2222 BC

Hvar Pottery

Petric [23] has considered the late Neolithic pottery of the Hvar culture (which is located on the island of Hvar in the Adriatic sea) and reached the conclusion that a major event occurred in the skies at the end of the Fourth Millennium BC that was reflected in a major cultural change, at least so far as the motifs found on the pottery are concerned.

The Köfels object would make a spectacular impression on the people of the Adriatic. The pre-entry phase would be visible as a brighter object than seen by the Sumerians. It would disappear in the earth shadow before reappearing close to the Moon in the southeast. The trajectory passes directly over Hvar itself when it would appear as a fireball four times larger than the sun. The later plume would be equally spectacular, rising from the northwest and then arching over Adriatic observers towards the southeast.

The examples of pottery Petric gives are suggestive rather than conclusive and show many different variations. However eyewitness descriptions of the Tunguska event have a wide range of visual interpretations from column, to wedge, to split in the sky [24] that parallel the pottery images.

A particularly intriguing pot fragment (Figure 20) shows a V shape close to the Moon. Figure 21 shows the trajectory of the

Figure 20: Hvar Pot Shard (Petric [23])

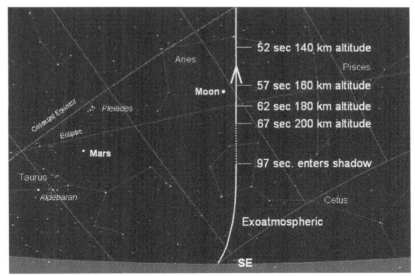

Figure 21: Object Track from Hvar Close to Moon *(Times are to Impact)*

asteroid as seen from Hvar as it passes close to the Moon and it can be seen that the pottery image is consistent with the asteroid at around 150 km altitude. This suggests that the supersonic shock wave can be seen, marked out by the ionised air, before the object itself becomes luminous due to heating. This detail of the dynamics of large bodies entering the atmosphere has not been noted before.

Petric dates this pottery to around or just before 3000 BC, though this dating of the Hvar culture is still a matter of debate, with other experts suggesting dates hundreds of years earlier. However the later Petric dating is viable and consistent with a 3123 BC event.

Climate Change

It is possible an event of this magnitude could cause a discernable climate change, and there is evidence of an abrupt and major climate change towards the end of the third Millennium with both earth science and archaeological evidence of major culture change. This change has been widely discussed e.g. Dalfes *et al* [25]. So the 2222 BC date would be consistent with this change. However there is also evidence of an abrupt and major climate change around 3100 BC, known as the Piora Oscillation, although discussion of this period is less extensive and focused.

Again there is earth science data (e.g. tree ring changes [26]) and archaeological evidence such as the disappearance of the late Chalcolithic culture in the Levant, the unification of Egypt, and the end of the Uruk period in Sumer.

So climate change evidence cannot be used as a means to discriminate between the two dates. However once a final date is established then it may cast light on the reasons for one of the Bronze Age collapses.

9.3 DATING CONCLUSIONS

Given the interpretation of the two planets in Cancer and a planet in Gemini only and without an assumption about a planet in Taurus, then 18 years in the timeframe 3500BC to 2000 BC were found to meet this condition on the 20th day in the midsummer month. Half the matches were three planet matches and half also matched a planet in Taurus. None of the three planet matches were of good quality and this gives further credence to the identification of the object in Taurus as a fourth planet.

Of the three dates that are judged to give good visual matches, 6 July 2269 BC has been ruled out because of a very poor trajectory match with Köfels and it is argued that the great preponderance of evidence suggests a date of 29 June 3123 BC rather than 29 June 2222 BC. However it is acknowledged this is not absolutely conclusive.

ANALYSIS OF THE ORBIT, TRAJECTORY AND PLUME

10.1 TRAJECTORY

The track across the sky defined by the Path Sector together with the precise timing enables the trajectory of the Köfels object to be accurately determined. An analysis has been performed, using the data available on K8538, and with the objective of checking the credibility of the hypothesis that the observation on K8538 is related to the Köfels site. From the best fit trajectory, the original heliocentric orbit is found which gives considerable information about the nature of the body involved. This chapter also considers the dynamics of the back plume.

The trajectory analysis is entirely dependent on two sources of information: the suspected impact site at Köfels and the data on tablet K8538. The impact site in the Austrian Alps has precisely known latitude, longitude and altitude but with a much less precise estimate of impact energy and bolide approach azimuth. Tablet K8538 contains specific astronomical data relating to time, visible appearance and path of the object against the stars. Unfortunately the exact location of the observer who recorded the data on K8538 is not known.

The first task was to narrow down the position of the astronomer (a justified epithet since K8538 was clearly produced by a professional observer) using internal evidence in the record. The tablet contains the following information.

    i.   The time from the start of the Sumerian month.

    ii.   The time from last sunset and the anticipated time to sunrise.

    iii. The size and appearance of the body.

    iv. The approximate path of the body through the constellations.

v.   The exact path of the body through Pegasus & Aquila in relation to Enif & Altair.

vi.   The constellations visible just prior to dawn, fixing the time of year.

vii.   The location of the Celestial Equator is shown.

viii.   The position of several unnamed planets.

Items 1, 2, 6 and 7 above have been analysed in Chapter 4 which discussed the observation errors to give a 'best estimate' of the location of the astronomer and, taking into consideration the known progress of urbanisation at the date inferred, the following analysis has assumed the location to be close to Kish in Southern Mesopotamia. This cannot be 100% certain, but adoption of the likely alternatives does not significantly alter the conclusions.

Next, an approximate date, time of year and time of day was needed. This is discussed in Chapter 8. The date was determined based on the planetary configurations given on K8538 and narrowed down to 29 June 3123 BC being the strongly preferred date and the one used for all analysis.

Scale can be given to the impact by examination of the Köfels feature. The geology at the Köfels site will be discussed in Chapter 12. It was concluded using data for nuclear weapons effects [28] that the impact energy is estimated to be approximately 10^{10} tonnes of TNT equivalent (4×10^{19} Joules) based on the approximate crater radius in hard dry rock. The errors on this calculation are hard to estimate, but the energy released could credibly be within a factor of two of this value.

Using evidence solely from the impact site the direction from which the object came is not well determined, it being only possible to say that it arrived from the eastern hemisphere having 'struck' the eastern side of the mountains.

There were four primary software tools for the trajectory analysis outlined below.

i)   A non-commercial trajectory program, Bolide3, which uses a spherical rotating Earth with the US1962 atmosphere model. The program employs a modified Euler integration scheme. It does not have higher harmonics of the gravitational field, nor perturbations for the Sun and Moon. The program is derived from software originally written by one of the authors for modelling launch vehicle and ballistic missile trajectories. The program has an approximate ablation model and has the starting mass, density, shape factors and aerodynamic coefficients as input. Bolide3 entry trajectories are modelled in reverse, starting at the terminal location with a trajectory flight path angle, azimuth and velocity, the signs of which are all reversed (including the Earth's rotation) to model the arrival as a departure trajectory.

ii) A non-commercial program Phaeton3, which was purpose written, simulates the appearance of a Bolide3 trajectory from any specified location around the Mediterranean, North Africa, the Middle East and Europe. The program outputs a polar plot of the trajectory as altitude - azimuth relative to the observer. The program also calculates the angular size and illuminated phase of a sphere of stated diameter on the flight path taking account of the location of the Sun, which is also part of the input. Example screen shots are shown in Figures 22 and 23.

iii) A non-commercial program Radiant5, written to study bolide origins. This is a simple patched conic sections program that converts the hyperbolic Earth centred encounter (as determined by Bolide3) into its prior Helio-centric elliptical or hyperbolic orbit, as the case may be.

iv) The commercial astronomy program Redshift 4 and 5 (discussed in Chapter 2). These programs enable an accurate determination of the appropriate astronomical background for the dates considered and can derive altitude azimuth data for the relevant stars Enif and Altair. Care has to be exercised with this program near the horizon where the effects of refraction are only approximately allowed for by simply depressing the horizon.

The relation between the programs is shown in Figure 24. Using this suite of software extensive simulations were performed with systematic variation of the input parameters in Bolide 3 to see if the data on K8538 could be reproduced. Specifically, the effects of mass, density and the terminal velocity, flight path and azimuth were investigated.

The trajectory was tracked back to a position a few thousand km above the Earth's surface. The aerodynamic drag had a noticeable effect on the terminal velocity and the density of the body could

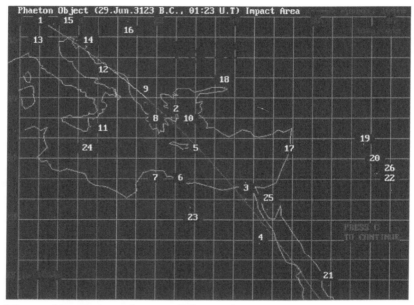

Figure 22: Starting Screen of Phaeton3
showing locations of selected sites around the Mediterranean and the ground track of the Kofels object colour coded for altitude (0-100km, 100-200km, 200-1000km, >1000km).

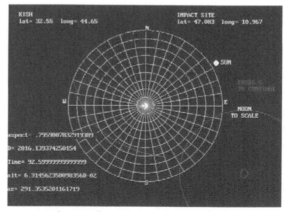

Figure 23: Example Display Screen of Phaeton3
in this instance from Kish (site 26 in figure 22) showing the path across the sky and, inset to the right, the appearance of the body relative to the Moon for scale. The Sun is below the horizon. The panel to the left displays the apparent aspect ratio, line of sight distance, time from impact, altitude and azimuth at that time.

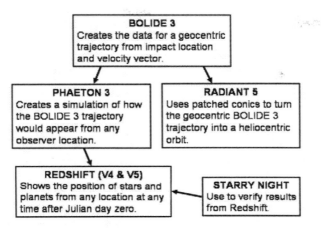

Figure 24: The Trajectory Analysis Software Suite

therefore be varied to produce significant variations of the trajectory. However, by far the most significant variables that were found to influence the observation from Mesopotamia were the flight path angle, the azimuth and the terminal velocity. Results for the 3123 BC date are presented in the Table 5.

There is an important item of supporting information in the Path Sector with the reference to 'large star' at a point along the detailed trajectory between Enif and Altair. This can be seen either as a gratuitous statement or, more reasonably, the position where the astronomer judged the body to have maximum apparent size.

Thus we know the path across the background stars and the approximate observers azimuth to the point of closest approach. This ties down the flight path angle and azimuth quite well to approximately -6° (±0.5°) flight path angle and approach to Köfels from an azimuth of 132° (±0.5°). These two parameters are linked when trying to maintain a reasonable fit to the path through Enif and Altair, but the point of closest approach is largely determined by the azimuth. Assuming the trajectory given by case 9 the azimuth of closest approach is 258.2°, about 39% of the separation from Enif to Altair. This is a very good fit with K8538. The ground track of Case 9 is shown in Figure 25.

K8538 shows the trajectory passing through the two stars, with perhaps a hint that it was slightly below Enif. This is quite

Table 5: Trajectory Simulations for 3123 BC

Case No.	Flight path angle (°)	Azimuth (°)	Terminal velocity (m/s)	Density kg/m³	Pass Enif (°)	Distance Altair (°)	Comments
1	6.0	131.0	11,700	600	1.81	-0.14	3:2 resonant orbit
2	6.5	131.0	11,700	600	1.07	0.38	
3	6.0	130.5	11,700	600	1.17	0.31	
4	6.75	131.0	11,700	600	0.71	0.66	equal error spread
5	6.0	131.5	14,000	600	0.37	0.51	
6	6.0	132.5	16,000	600	0.51	0.19	
7	6.0	133.0	18,000	600	0.42	0.13	max. credible speed
8	6.0	132.0	15,000	600	0.37	0.37	equal error spread
9	6.0	132.0	14,950	600	0.40	0.37	2:1 resonant orbit
10	6.0	132.0	14,950	1200	1.02	0.03	
11	6.0	132.0	16,050	1200	0.40	0.37	2:1 resonant orbit

unlikely to have occurred in practice. The lighting conditions at the time implied on K8538 would mean that only stars brighter than magnitude +3 would be readily visible. Enif is a magnitude 2.54 star and Altair magnitude 0.89. There are only 138 stars brighter than magnitude 3, or on average 11 per steradian of sky. In its passage across the sky the body (typically with an angular diameter of 3 arc miuntes) would occlude a total sky area of perhaps 2.3 x 10⁻³ steradians. The probability that it would therefore occlude any star above magnitude 3 is only about 1 in 40, i.e. about 40 such objects would need to pass by at random before a bright stellar occlusion would be expected to occur. The probability of 2 such occlusions by the same body is only 1 in 3200.

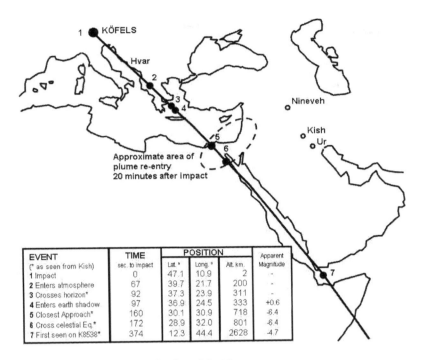

| EVENT | TIME | POSITION | | | Apparent |
(* as seen from Kish)	sec. to impact	Lat. °	Long. °	Alt. km.	Magnitude
1 Impact	0	47.1	10.9	2	-
2 Enters atmosphere	67	39.7	21.7	200	-
3 Crosses horizon*	92	37.3	23.9	311	-
4 Enters earth shadow	97	36.9	24.5	333	+0.6
5 Closest Approach*	160	30.1	30.9	718	-6.4
6 Cross celestial Eq.*	172	28.9	32.0	801	-6.4
7 First seen on K8538*	374	12.3	44.4	2628	-4.7

Figure 25: Ground Track of Köfels Object

However, as already discussed in Chapter 8, it is unlikely the observer was concerned with precise positions as he would not gain any meaningful value from the data. However the relation between objects would have been important so if it is assumed the tablet is indicating a very close approach then a very different level of probability is found.

For the purpose of trajectory comparison it was assumed that the body passed equally close to Enif and Altair. This was an arbitrary choice and other strategies could of course be employed for the comparison. It was found following many trajectory simulations that the miss distances could be minimised to about 0.4° (see Table 5). It is possible to estimate roughly how realistic it would have been for the observer to have seen two such close approachces. The path of the object in its visible portion covered about 150° of arc and for a star to be within ±0.4° of it would define an area of sky of 0.0366 steradians. The Poisson mean for a star to be in this area is then 0.402. The probability of no bright stars falling within this area is about 67%. Thus there is a reasonable chance that at least some stars will be passed at this distance. The probability of passing only one star is 27% while the probability of passing two is about 5% and three below 1%. The observation of a 0.4° close approach to Enif and Altair is therefore realistic, although at a probability of 1/20, fortuitous.

It is statistically significant that this improbable trajectory could be modelled accurately and consistent therefore with the hypothesis that K8538 could indeed be a record of an observation of the object that struck at Köfels, provided that it is accepted that the trajectory depicted is unlikely to be literally correct and the object simply passed very close to the stars and not literally occulted them.

10.2 HELIOCENTRIC ORBIT

Throughout the observation from Mesopotamia the body would have been exo-atmospheric, entry into the atmosphere only occurring when it was halfway along the eastern coast of the Adriatic Sea and below the horizon of the astronomer.

K8538 gives an approximate size for the body, stated as 2' by 3' of arc. This translates to a body approximately 1.37 km diameter at the calculated minimum range to the observer (about 1570 km) from the trajectory analysis. However the impact energy at Köfels is unlikely to be very far from 10^{10} tonnes of TNT equivalent. Hence the body would need to be of low density and relatively slow moving. The error in the size observation is unknown, but it is true that the object would be seen to be shortened due to the illumination phase, assuming it to be roughly spherical. The observation is however close to the limiting acuity for the unaided human eye, which can separate features subtending 0.2' to 1.0' depending on the nature of the field being viewed. Trials with a model 'asteroid' subtending this angle confirmed it to be a readily resolvable object, especially to a young eye.

There has been extensive study of the ballistic characteristics of bodies entering the Earth's atmosphere through the Prairie Network and European Network Fireball programs [29,30], although these bodies were obviously much smaller than that being discussed here. This work has provided information on apparent density and terminal disruption of these bodies. Adoption here of a density of

600 kg/m^3 would be consistent with a type IIIa bolide [30] and leads to a credible impact energy. This would also be consistent with a body that may well have already been undergoing break-up at the time of impact. Thus a body with a mass of approximately 6.0×10^{11} kg is proposed, although twice this value may be possible.

Having used Bolide3 to determine various trajectories constrained by the K8538 data, the terminal velocity was varied in order to derive exo-atmospheric state vectors for the object at about 4,000km altitude. This output was then input into a program, Radiant5.

All the proposed trajectories led to an object in an Aten-like orbit for the likely range of impact velocities (12 to 18 km/s). The more interesting of these are shown in Table 6. The lower velocities were more in keeping with the likely impact energy at Köfels for a body with a mass of 6×10^8 to 12×10^8 tonnes, while higher velocities gave a better match to K8538. Another feature which was noticed during this analysis was that the line of apsides for the Heliocentric orbit fell very close to the plane of the ecliptic, while the aphelion was close to the radius of the Earth's orbit. A terminal velocity of 11.7 km/s gave a body, which was close to an orbit with a 3:2 resonance with the Earth's orbit, while a terminal velocity of 14.95 km/s gave a 2:1 resonant orbit. This, together with the two factors above were considered to be too great a coincidence and we propose that the body was in fact driven into a 2:1 resonant capture orbit by gravitational interaction with the Earth over some period of time. This was Case 9 in Table 5 above.

It can be seen that while the characteristics of Case 9 are broadly those of the Aten group (Table 7) there are significant differences in the parameters relative to those of the 'mean Aten'. Assuming that this is the result of interaction with the Earth it is possible to

Table 6: Helio-centric Orbital Parameters

Case No.	semi-major axis a (AU)	perihelion q (AU)	eccen-tricity e	orbit inclination i (deg)	long. of ascending node Ω (deg)	argument of perihelion ω from Ω (deg)
1	0.7565	0.5028	0.3353	11.2708	319.616	177.684
7	0.5784	0.1412	0.7558	23.9915	319.618	176.482
9	0.6303	0.2483	0.6061	17.186	319.617	176.665
11	0.6293	0.2463	0.6086	17.2494	319.617	176.639

Table 7: Typical Aten Asteroid Characteristics

Aten Identifier	a (AU)	q (AU)	i (deg)	e	radius (km)
2062	0.9665	0.7900	18.93	0.1825	0.4
2100	0.8321	0.4689	15.76	0.4365	0.8
2340	0.8439	0.4643	5.856	0.4497	0.1
3362	0.9897	0.5260	9.923	0.4686	0.7
mean Aten	0.9081	0.5623	12.6172	0.3843	0.5
Case 9	0.63	0.25	17.19	0.606	1.2

carry out a simple calculation to check that the mean Aten can be transformed into Case 9.

Ignoring the relative orbit inclinations of the two bodies, it is assumed that the Earth overtakes a 'mean Aten' and during the encounter transforms its velocity vector so that the asteroid leaves the Earth's vicinity with its velocity vector parallel to that of the Earth, but now travelling more slowly. Relative to the Earth the asteroid would approach from deep space at 11km/s ahead of the Earth and at an angle of 18.8° to the radius vector to the Sun. Following the encounter the asteroid would leave the Earth with the same relative velocity but now parallel to the Earth's motion.

This hypothetical asteroid's post encounter orbital parameters are calculated to be a = 0.624 AU, q = 0.248 AU and e = 0.603. The comparison with Case 9 is remarkable and must contain a certain amount of coincidence. A calculation was performed to assess whether such an orbit transformation could be achieved in a single encounter. However, this would require the asteroid to pass within 2364 km of the Earth's centre and so is not possible. It is however very clear that the Earth could have undergone several near encounters with the asteroid, gradually reducing its aphelion, perihelion and semi-major axis until it became close to resonant and an impact was inevitable.

Assuming the body to have been one of the Aten group, it would be expected to have a low albedo since these objects are hypothesised to be volatilised comet remains, also consistent with the low density postulated above. Here we take the albedo to be 0.05, which may be slightly pessimistic. It is possible to calculate the apparent brightness of the object in order to check the Sumerian astronomer's ability to see it and make the

Table 8: Visual Magnitude of Body from Mesopotamia

Time Before Impact (sec)	Distance from Observer (km)	Altitude (degrees)	Visual Magnitude (corrected for air absorption)
890	11,468	24.96	-2.31
622	7,461	28.06	-3.24
426	4,510	32.08	-4.33
344	3,311	34.57	-5.01
278	2,422	36.32	-5.68
236	1,949	35.67	-6.06
196	1,647	30.71	-6.32
170	1,572	24.10	-6.33
122	1,738	8.60	-5.41
93	2,016	0.0	+1.35

observations recorded. The results are presented in Table 8, assuming the observer's assessment of the angular dimensions to be correct. For comparison, Venus when bright has a magnitude of about -4.4, so that this object would have been much brighter than Venus throughout most of its observed trajectory on K5838.

10.3 TERMINAL ENTRY

Entry into the Earth's atmosphere is a vague concept because there is no finite altitude at which the atmosphere can be considered to end and the vacuum of space begin. Further, the density at high altitudes varies on time scales ranging from hours to years depending on the motion of the Earth and variations in the Sun's radiant output and solar wind.

For practical calculations on trajectories it has become convention to begin atmospheric entry at 122 km (400,000 feet) which encompasses the range over which aerodynamic forces can have any significant effect on the motion of the body. However, analysis of the Tungus object by one of the authors has suggested that the object was showing visible interaction with the atmosphere to greater altitudes, possibly 190 – 200 km. At much lower altitudes, 70 km and less, the entry is clearly in progress and a strong shock wave is radiated from a body travelling at hypersonic speeds.

Evidence from the Hvar pottery discussed in Chapter 9 also suggests the object could be clearly seen at around 150 km altitude. This seems to show shock heating of the air would dissociate it and partially ionise it at a temperature of about 12,000K, emitting light by recombination in the gas in the bow shock and the wake shock of the body.

For the body considered here the trajectory would have followed closely the eastern coastline of the Adriatic sea and the conventional entry would have begun above the location of Dubrovnik about 45 seconds before impact. At 40 seconds to impact the object would have fallen to 100 km altitude approximately above Hvar. From Dubrovnik the object would have subtended an angle of about 39', or slightly larger than the Moon, growing rapidly larger as it increased its latitude.

The calculated energy deposition into the atmosphere during entry is shown in Figure 26. This reaches very high levels, amounting to the equivalent of 'atom bombs per metre', and the implied shock strength reaching the surface of the Earth would have resulted in widespread destruction at the north end of the Adriatic, across Istra and the Venetian plain into the Southern Alps.

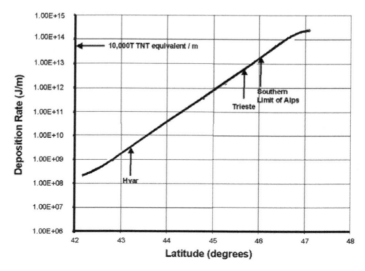

Figure 26:   Energy Deposition Rate Vs Latitude

At the Southern Alps the over-pressure on the ground would reach several atmospheres, while just prior to impact the blast wave pressure on the ground would be around 100 atmospheres. This local pressure would have been greatly exceeded in the explosion following the impact.

An object of this size would normally be expected to produce a significant crater. It is clear from the absence of such a conventional crater that the object had exploded before it reached Köfels.

The details of break-up of large bodies entering the Earth's atmosphere is only vaguely understood. It is known that smaller carbonaceous bodies particularly those with a shallow trajectory can explode producing a bolide; the largest known example being the Tunguska event in 1908. Comet Shoemaker-Levy 9 showed large bodies do airburst on Jupiter and there is some evidence that very large bolides can explode close to the ground and create geological structures on Venus [31], but both these planets have thicker atmospheres than the Earth so the degree to which this applies to the Earth is currently a matter of debate [32, 33, 34].

However this debate on the likelihood of airburst of kilometre-sized objects is irrelevant in the case of the Köfels impact. The trajectory determined from K8538 allows a detailed understanding of the final impact dynamics. It is an inevitable consequence of the shallow trajectory that there is a complex interaction of the object and ground features that spreads the impact over around 20 km. To understand these processes requires an understanding of the site and the dynamics of the impact are considered in detail with the other site discussion in Chapter 12.

10.4 PLUME

The impact of Comet Shoemaker-Levy 9 on Jupiter in July 1994 showed the importance of the plume which is ejected from an impact roughly back in the direction of the incoming trajectory plane [35]. Although the plume had been predicted, the impact on Jupiter showed that the models are far from accurate when the detailed structure is considered. It is not possible even to predict the energy of the plume and the fact that all the plumes on Jupiter reached the same altitude is not adequately explained.

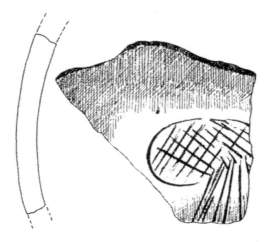

Figure 27: Plume Illustration on Hvar Pottery (Petric [23])

By analogy we expect that the impact at Köfels would have produced a plume. This may have carried away some 10% of the impact energy and possibly up to 70% of the mass of the object.

The early "mushroom cloud" phase of the plume would have been self illuminating and judging by one pottery fragment (Figure 27) could be seen from Hvar.

The initial plume velocity would be expected to be 5 km/s and to have an angle of around 55° - caused by the shallow angle of the asteroid's entry. It would have passed across the Mediterranean Sea back basically along the incoming path but shifted eastward by coriolis forces (any lift forces generated by the asteroid would also have affected the trajectory). The rough modelling of the plume suggests the reentry would be over the Levant, Sinai and northern Egypt, possibly reaching into modern Syria.

While over the Mediterranean, it would reach an altitude of over 900km. This would be sufficient to be ejected into the sunlight and to have been observed from Mesopotamia. The tablet K8538 may indicate that the plume was observed as a "tree-like phenomenon". Simple ballistic trajectories launched from Köfels show that the plume may have reached an altitude of approximately 20° at an azimuth of 290° viewed from Kish. The rapid approach of dawn at Kish would have limited the time for viewing the spectacle and it was probably not visible throughout its flight.

The energy dissipated in the atmosphere on re-entry of the plume would be approximately 4×10^{18} J, which if spread over an area of up to one million square km, would be sufficient to cause conflagration of any exposed combustible material, including people. It is probable many more people died under the plume than died in the Alps due to the primary impact.

Given a considerable fraction of the asteroid material would be contained in the plume, there would also be significant contamination of the area where the plume re-entered which should produce an iridium concentration in the order of 10^{-7} kg/m^2.

It should be noted that the analysis of the plume is rather crude. A more refined analysis of the plume dynamics was beyond the current capabilities of this study.

10.5 ANALYSIS CONCLUSIONS

Extensive trajectory analysis has been performed using computer programs to model the events depicted on the Akkadian astronomical tablet K8538 showing that they are linked to the object which impacted at Köfels in the Austrian Alps. A number of trajectories that fit the observation have been explored and the 'best fit' trajectory was found to be:

$V_{terminal}$ = 14.95 km/s   Azimuth = 132°   Flight path angle = 6°

From the variation in the different case trajectories (Table 5) and from consideration of the ground site evidence (outlined in Chapter 12), the Azimuth is thought to be accurate to one degree and the Flight Path Angle to half a degree. The velocity could have a much wider range from 11.5 km/sec to 18 km/sec.

The "best fit" trajectory matches the following heliocentric orbit which is 2:1 resonant with the earth's orbit.

a = 0.63 AU   q = 0.25 AU   e = 0.606
i = 17.19°   Ω = 319.62°   ω = 176.67°

As with the terminal conditions, the accuracy with which the results are shown is to allow them to be checked. However the fact that a resonant orbit comes out of the analysis, a result extremely unlikely to come from random parameters, is strongly indicative that the accuracy is good to two significant figures for the "best fit" trajectory.

As Table 5 shows, other trajectories within the accuracy argued for the terminal conditions also yield resonant orbits. From this it is concluded the object that impacted at Köfels, on 29 June 3123 BC, was an Aten asteroid that had been driven into resonance with the Earth's orbit prior to impact.

11

THE PLUME AND NORTH SECTORS

11.1 PLUME SECTOR

The remaining two sectors of K8538 are largely missing but are consistent with observations of post impact phenomena.

The Plume Sector (Figure 28) is highly damaged but the ends of three lines of text are visible running radially towards the centre of the tablet from the rim. The upper line reads:

SUD (B373) SUD (B373)

which translates as "very distant."

The second line simply shows the end of a cuneiform sign, which could be B339, B401, B548 or B555. It is suggested that GIBIL$_2$ (B548) "burning" would be consistent with the appearance of the vast plume illuminated by sunlight believed to be portrayed here.

The third sequence shows text at the end of a line: %%% ÑIŠ (B296) ÑIŠ (B296) "a naturally branching tree-like phenomenon" indicating the shape of the plume.

As shown in Chapter 10 a very large plume would be expected, reaching an altitude of over 900km and arching over the Mediterranean Sea. Although this sector is very badly damaged what remains is consistent with the depiction of this phenomenon.

The rough modelling of the plume trajectory suggests it rose some 20 degrees above the horizon and could be seen starting in western Hercules, clipping Sagitta and Vulpecula before descending into east Aquila. So this interpretation of the sector follows the succession around the sky used in the remainder of the tablet.

11.2 NORTH SECTOR

A significant proportion of the North sector (Figure 29) is damaged and therefore it contains some ambiguity.

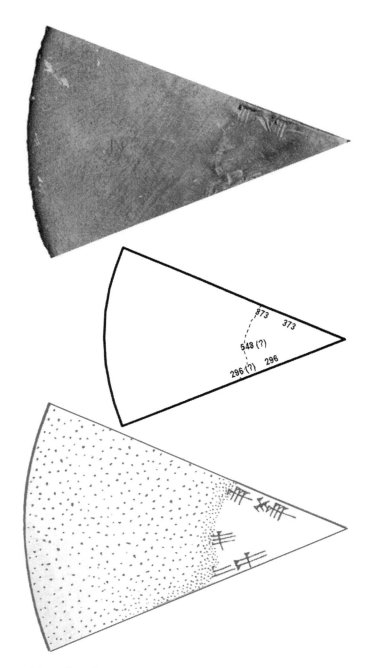

Figure 28: The Plume Sector

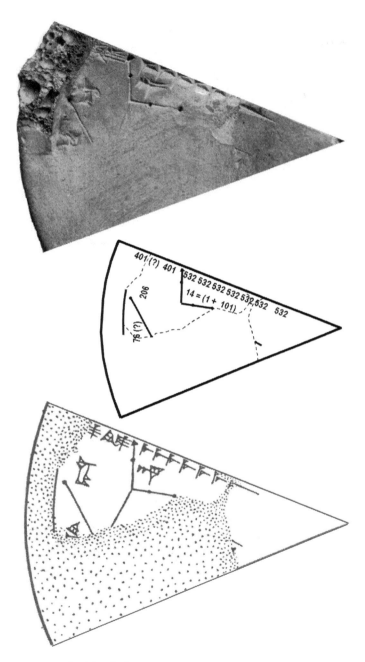

Figure 29: The North Sector

That it is the North sector is first suggested by its relative position to the other sectors but also by the sign GUB (B206) "stand" next to a dot. The obvious interpretation of a star that stands still is the pole star, which for the Sumerian period is Thuban. It should be noted that there was no significant pole star in the First Millennium when K8538 was produced, indicating a much earlier date of Third or Fourth Millennium BC for the original document. This proposition is reinforced by the position of the Celestial Equator in the Pisces sector which could not have been later than the Third Millennium BC.

A good match has been found with stars from the modern constellation Ursa Major (Figure 30), except for the object represented by a triangular impression on the upper boundary. It is suggested that this object is actually Capella in Auriga. The direction is correct but it is significantly further than indicated and well outside this sector. The reason Capella might be an important reference star is that it is located very close to the Celestial longitude of the spring equinox. A Redshift 4 simulation of the North Sector of the sky is shown in Figure 31.

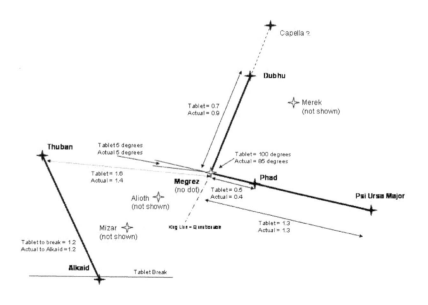

Figure 30: Comparison of North Sector Drawing with Presumed Real Sky

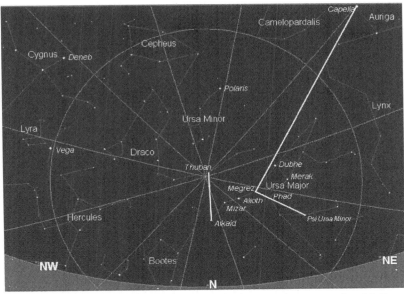

Figure 31: Redshift 4 Simulation of The North Sector for 29 June 3123BC. *The construction in Figure 30 is shown superimposed*

It should be noted that Megrez, a somewhat dimmer star than others in the body of the Bear and is not shown as a dot. This provides further evidence dots are only used when a star is visible, but also the observer is prepared to indicate stars he cannot see.

This identification would suggest that the progression across the sky has been followed. It seems to cover a large area of the sky and brings the tablet full circle back to the Cancer sector.

Along the upper boundary is the sequence;

%%% MUR (B401)? MUR (B401) ME (B532) ME (B532) ME
(B532) ME (B532) ME (B532) ME (B532) ME (B532)

Possibly "region {dark, .. , ... , .. , .. , .. , .. }". The term for region was used in the Cancer sector but the use of ME is problematical unless it is assumed actually to be the homophone ME_2 (B427), also pronounced MI "dark".

The solitary symbol AŠŠUR (B14) appears in the region thought to be Ursa Major. In the past it has been assumed this relates to the deity Assur. However this is an Akkadian word and it seems more probable that, following Labat [13], it is the two Sumerian signs AŠ (B1) SUR (B101) "the one who bounds" run together as found elsewhere. This would be a good description of Ursa Major, which circles the edge of the polar region. The stars shown are thought to be Phad linked to psi Ursa Major, and Dubhe linked to Capella. Psi Ursa Major always remains just above the horizon in this period from Sumer.

Emerging from the damage is the end of a sign that could be GE (B85), MUD (B81) or MAŠ$_2$ (B76). MAŠ$_2$ "the he-goat" would make sense since a constellation of this appellation is described in the MUL APIN text as being in the vicinity of modern Hercules [6]. However this interpretation must now be questioned as the scaling of the picture in the sector shows (Figure 30) that it can extend no further than Bootes.

Figure 30 also shows that like the Cancer sector the lower edge may have been the horizon but since the stars drawn are rather cramped, this may also explain the inaccurate angle from Megrez. In all previous sectors where star maps have been drawn their purpose has been to locate some transitory event, but it also clear the missing sector contains the key subject of interest and the stars left on the tablet are there to show the boundaries of the feature.

From the comet Shoemaker Levy 9 impact on Jupiter it is known that strong aurora effects accompany large body impacts [36]. These would be visible from Sumer in the Northern sky and are possibly what is being recorded here. It would almost certainly be the first time the Sumerian observers would have seen aurora which would heighten the sense of strangeness. The "dark" labels remaining are indicating regions free of the lights in the northern sky.

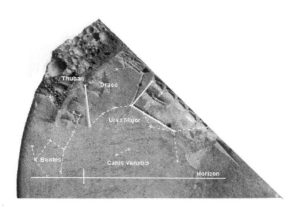

Figure 32: North Star Field Superimposed on North Sector

12

THE EVIDENCE AT THE KÖFELS' SITE

12.1 SITE OVERVIEW

The Köfels geological structure is around 5 km in diameter and named after the small village close to the centre of the feature in the valley of the Otztaler-Ache in the Austrian Tyrol. The main geomorphic feature is a massive landslide on the western side of the valley, which filled the valley producing the Maurach barrier. This extends to the other side of the valley. The landslide depth was around 500 m.

Since the landslide occurred, the river Otztaler-Ache has cut a smaller valley through the barrier. This secondary valley has not yet reached the base of the landslide material.

The site country rock is granitic gneiss with an age around 84 Million years [37], but the petrography also includes a unique "pumice" like rock called "Köfelsite" which forms dykes within the feature. Where the dykes meet the surface the cooled material forms a glassy layer. All researchers agree this "Köfelsite" was formed at the same time as the landslide process.

12.2 ORIGIN THEORIES

The site was originally discussed as a volcanic feature by Pichler [38], but this was questioned by Suess [39] and Stutzer [40], who both independently proposed a meteorite impact as the explanation for the site's features. A detailed site description from Kranz [41], with this explanation in mind, highlighted that the feature must be after the last glacial event, that is within the last 10,000 years. Storzer *et al* [37] also concluded that the temperatures and pressures required to produce the "Köfelsite" are not possible with volcanic activity and a meteorite impact was the likely explanation.

These arguments have led to Köfels being identified as a possible impact site by the British Museum meteorite catalogue [42] and it was also included in a review of possible impact sites by Shorts and Bunch [43]. The key feature cited as pointing to an impact origin, is the presence of shock metamorphism in quartz crystals in both the Köfelsite itself and the country rock immediately adjacent to the Köfelsite dykes.

Two papers, one by Heuberger *et al* [44] and the other by Masch *et al* [45], reviewed the Köfels site in comparison with a landslide event in Nepal and concluded the rock melting was part of the landside process itself rather than due to the cause of the landslide. This initiated a more recent body of thinking that argues the Köfels site features can be explained by normal landslide processes. Erismann *et al* [46] have shown experimentally that the Köfelsite could be generated in the landslide. Lerouz and Doukan [47] have argued that the embedded shock quartz found within the Köfelsite is also a feature consistent with normal landslide events.

The current body of expert opinion believes that the melted Köfelsite rock and the shock metamorphism in the quartz was due to the landslide process and - with one caveat regarding the possible two stage development of the Quartz shock metamorphism to be discussed below - this is accepted in interpretation of the evidence argued in this book. However none of this evidence enables a clear cause of the landslide to be identified and it certainly does not preclude a NEO event being the initiator.

It should also be born in mind that in his site description Kranz [41] includes other evidence that argues against a simple landslide. He reports large granitic gneiss boulders scattered over two kilometres east of the landslide and at altitudes above the landslide and where no normal geo-morphological process could place it. He was also puzzled by the volume and location of "blasted rock (a mixture of boulders, sand and rubble)". Kranz concluded a massive explosion must have accompanied the landslide, locating it at the base of the western side of the valley. Since volcanism can be ruled out a NEO impact remains the only probable cause and is therefore consistent with the conclusions drawn from the Early Bronze Age observations.

While Kranz's work was conducted well before the current understanding of NEO impact processes was established, it still has the advantage of taking a holistic view and the evidence he cites requires explanation, which later more focused work still has not provided.

12.3 THE IMPACT

There is no clear crater at Köfels, which implies mid-air explosive destruction of the object before it reached the Köfels. To create the pressures and forces required to initiate the landslide, the explosion would need to very close to the ground. Chapter 10 has discussed the general debate on airbursts. However the trajectory of the object as determined by K8538 means that special circumstances apply and the general debate has little relevance. Judging from the possible secondary impact site at Felderkogel (which is a traditional crater as discussed in Section 12.6), the secondary object creating it must have been solid, suggesting the main object would also have been solid when it reached the ridge ending with Gamskogel ridge.

The incoming path of the trajectory over approximately the last second of its flight is shown in Figure 33. It can be seen that the Köfels object flew almost perfectly parallel with and slightly to the east of the ridge ending in Gamskogel. Figure 34 shows the corresponding profile of the terrain over the last 15 km of flight and it is clear the object interacted with Gamskogel clipping the ridge to give a 2 kilometre long cut with a 6 degree slope, matching the incoming trajectory. This is clear evidence for physical contact with ground before reaching Köfels.

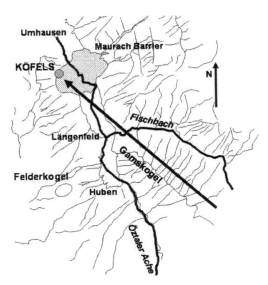

Figure 33:   Ground Track of Trajectory for Last Second

The Gamskogel interaction may have been a glancing physical hit or just the shock wave reflecting back on to the object (Figure 26 shows that energy in the shock is of the order 10^{14} J/m^2). Either way it has clearly sculpted the Gamskogel ridge (Figure 34 and Figure 35) and would undoubtedly cause the break up of the object. Thus when it left Gamskogel the object would no longer be solid and was either mostly or completely gaseous.

The object would have become an exploding fireball as it travelled down the Otztaler-Ache valley, which we estimate would

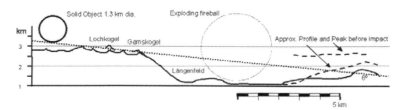

Figure 34: Terrain Profile along Flight Path

Figure 35: Gamskogel as Seen from Köfels. The Trajectory Centre and 1.3 km Diameter Circle Superimposed

be in the order of 5 km diameter at the time it struck the Köfels area. Thus the area affected was spread over several square kilometres and over a third of a second in time and with no structural strength in the impactor - hence no clear crater was formed.

If the object is diffuse without structural strength it could no longer support an external shock front to deflect the air ahead of it. This air would be entrained into the expanding gas cloud providing the energy to power the expansion. Although the object would be exploding due to the Gamskogel interaction, the entrained air would add energy exponentially and dominate the explosion process. This has been simplistically modelled, ignoring gas thermodynamic processes as shown in Figure 36.

The model calculates the mass increase due to the entrapped air and the corresponding reduction in velocity to conserve momentum. The reduction in speed means a loss of kinetic energy, which it is assumed becomes the kinetic energy of the expansion. The model numerically integrates over time to give the radius, mass and velocity history.

This model was run for an object 800 Mtonnes and a density of 600 kg/m³ with an initial velocity of 14 km/sec. In the 800 milliseconds it takes to cross the 11 km between Gamskogel and

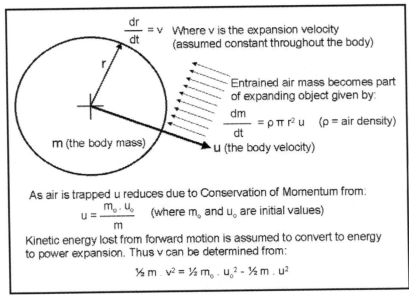

$$\frac{dr}{dt} = v \quad \text{Where v is the expansion velocity (assumed constant throughout the body)}$$

Entrained air mass becomes part of expanding object given by:

$$\frac{dm}{dt} = \rho \pi r^2 u \quad (\rho = \text{air density})$$

m (the body mass)

u (the body velocity)

As air is trapped u reduces due to Conservation of Momentum from:

$$u = \frac{m_o \cdot u_o}{m} \quad \text{(where } m_o \text{ and } u_o \text{ are initial values)}$$

Kinetic energy lost from forward motion is assumed to convert to energy to power expansion. Thus v can be determined from:

$$\tfrac{1}{2} m \cdot v^2 = \tfrac{1}{2} m_o \cdot u_o^2 - \tfrac{1}{2} m \cdot u^2$$

Figure 36: Air Entrapment Model Assumptions

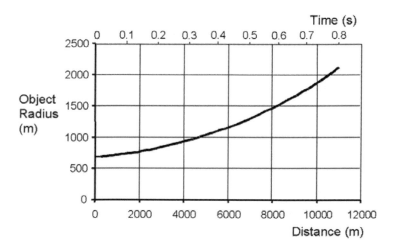

Figure 37:   Model Predictions for Growth in Radius.

Köfels it entraps 55 Mtonnes of air and slows to 13 km/sec. Figure 37 show the radius history as predicted by the model for an object showing it grows from 1.4 km diameter to 4.2 km.

Although a very crude model, it does produce some important conclusions that would not be altered by more precise modelling, as the exponential function rapidly takes the results into absurd regions. To get a fireball in the order of 5 km at Köfels (the size of the landslide) from an object in the order 1 km – 1.5 km (the size consistent with both the other site evidence and the size information on Tablet K8538) it must have a low density; something like 600 kg/m^3 to 1000 kg/m^3. An object with a higher density would need to be larger to reach 5 km due to the ratio of initial mass and entrapped air mass but this would raise the overall mass and hence the overall energy of the impact to unrealistic levels.

A density of 1000 kg/m^3 is consistent with the general evidence on the density of carboneous chondrite class asteroids [48].

The energy for the impact of 8.0 x 10^{19} J is four orders of magnitude higher than the previous estimate from Erismann et al [46] but this estimate was a minimum if a ground impact occurred to match the energy in the landslide.

The impact at Köfels has a momentum of 10^{16} kg m / sec being applied for 380 miliseconds. Spread over a 5 km diameter disk this gives a pressure of around 1.5 GPa. This is a rough calculation and assumes a perfectly uniform expanding object. Although it is possible that higher values could be achieved, particularly the forward shock front, nevertheless it is difficult to see how pressures could approach 10 GPa and this is well short of the 50 GPa that would be required to produce clear shock metamorphism in the country gneiss [49].

Although there were early reports of shock structures in the rock at Köfels [50], in later work Leroux and Doukhan [47] conclude there were no impact shock metamorphism features. However in examining the quartz in the gneiss close (within 20mm) to the dyke structures Leroux and Doukhan reported deformation features consistent with the fracturing of the rock as the dyke was being formed.

Leroux and Doukhan noted that there seemed to be a two stage process in generating the deformation features with a moderate temperature brittle phase followed by a higher temperature phase. The first stage could be due to the pressure wave from the explosion or alternatively it could be from the fracturing process as the slide plane was created, or again a combination of the two. Therefore these results do not absolutely rule out some form of pressure wave induced metamorphism caused by the impact in addition to the second processes of metamorphism that occurred as a result of processes within the landslide.

12.4 STORZER DATING

As already noted all references agree the Köfels event must have happened after the end of the last ice age. This means that human beings must have witnessed the event since Europe has been occupied for all of that time.

The earliest published work attempting to date the feature using physical (as opposed to geological) methods is by Storzer et al [37] using the analysis of fission tracks. The team examined two samples of the melted rocks. The first was a surface sample of the melted surface material and this gave an age of 3100 +/- 1200 years (1σ). The second sample was of the pumice like Köfelsite and, after considerable processing, this gave an age of 8000 +/- 6000 years (1σ) the very large error being due to very small sample of melted rock left after the processing. These dates were then compared to a single pine tree that was radiocarbon dated at 8710 +/- 150 years ago, although the complete basis for this radiocarbon date is not published. Storzer *et al* reached the conclusion that the 8000 year age was correct, although the error was much larger, and that the more accurate 3100 year age was due to a hypothetical later surface "annealing" event.

Storzer *et al* concede the fission track technique was being used at its "methodical limits", which explains why such credence was given to the radiocarbon date essentially over riding the fission track dates. However the conclusion has a logical flaw. The paper concluded from the heating effects on the rock that the Köfel event was a NEO impact, but if that were the case the same heating would ensure all living material on the surface would be vaporized and therefore not be available for dating. The fact that the tree remains were in a condition to be radiocarbon dated means they must have already been buried at some depth before the event and therefore were of a considerably older date.

Later work by Lang *et al* [51] on landslide dating specifically warns of the dangers of radiocarbon dating embedded material alone as this can be the result of an earlier burial process. So dates obtained this way can only ever provide a maximum age for a landslide, not an absolute date. In order to reliably radiocarbon date a landslide, both post slide material and embedded material are needed. And as already noted to survive the event any carboneous datable material must have already been buried to some considerable depth

Lang *et al* also briefly discuss the Köfels age of 8710 years without referencing the source, but presumably it is Storzer. They cast some doubt on it by warning that many other Alpine landslides have been dated considerably older than their true ages.

If the Storzer *et al* fission track dates are considered without influence from the radiocarbon date, the "bad" sample has such a large error it is best ignored, which leaves the "good" sample and an age of 3100 +/- 1200 years (1σ). The younger half of this probability distribution can be ignored as such an event in classical, medieval, or modern times would be well recorded and in any case the geological dating is very unlikely to be that much in error. This leaves a Bronze Age date consistent with the dating established from the Planisphere,

12.5 KUBIK DATING

A more recent work dating the event has been published by Kubik *et al* [52, 53]. The papers are primarily about measuring ^{10}Be and ^{26}Al nuclide production in surface samples due to cosmic ray exposure. However they go beyond simply dating the event. Like Storzer *et al* they compare these to the radiocarbon date of embedded tree remains and give these dates greater weight than the method of measuring cosmogenic radio nuclide concentrations. They then go further and use this "known" date of the Köfels event to calibrate the nuclide production rate.

Measuring cosmogenic radio nuclide concentrations is a comparatively new technique used to date when geological surfaces are first exposed. It uses the nuclide production rate due to cosmic ray exposure. The production rate values used are supported both by experimental measurements e.g. by Dai [54] and theoretical modelling e.g. Masarik and Reedy [55]. While these values take full consideration of the variation of cosmic ray exposure due to the 22 year solar cycle, they do not take account of the minima in that cycle over the last millennium (i.e. the Dalton, Sporer, Wolf and Maunder minimas) which would significantly increase the cosmic ray exposure. What the Sun's behaviour was in previous millennia is not known and therefore the cosmogenic radio nuclide technique must be calibrated against known dates.

The key argument for the quoted accuracy of measuring ^{10}Be and ^{26}Al nuclide production has been the comparison of other techniques used to date the end of the Younger Dryas, a cold period which ended abruptly 11550 years ago. This has been dated by several different methods with high consistency and the cosmogenic radio nuclide concentration results match these.

Kubik *et al* clearly saw the Köfels landslide as another good candidate to support and refine the calibration of the technique. They trusted the radiocarbon dating of three embedded samples which when calibrated give an age of 9800 ± 100 years old relative to 1995 i.e. around 7800 BC. This was the assumed date for calibration purposes. The same arguments presented above against the use of the radiocarbon dating of embedded material alone also apply here. If the event was caused by a NEO object, then these tree remains must have already been deeply buried at the time of the event to survive the fireball. Even if the event was a normal geologic landslip, use of these dates alone is risky unless supported by confirmatory post-slide material dating, which is not the case.

Looking in detail at the results from both the published papers [52, 53], a total of nine samples from Köfels have been measured. Of these, six are taken from the eastern side of the feature, which is a surface that could be the original pre-slide surface displaced. Three were taken in the western side from areas that must be genuine freshly exposed surface and therefore, as acknowledged by Kubik *et al* [53], should give the best results. These western results differ very markedly from those on the eastern side. One of these was from a loose boulder and, since it might have moved since the landside event, was removed from the data set for calibration and the surface separately dated as 7600 ± 950 years old. A second and third sample would give an even younger date if also used as dating evidence, but both these were judged to have an undeterminable soil cover history and were simply removed from the calibration data set.

If the assumptions were reversed and the three western results are considered alone, and without special Köfels' factors including the recalibration of the production rates, then the age results would be more like 7000 ± 1000 years. This is not a good agreement with a Fourth Millennium BC date but it is within the two sigma error range.

There is a further concern about the accuracy of this technique and indeed any dating technique that uses the ratio of an element's isotopes as a starting assumption. The incoming body contains sufficient material to contaminate the site with enough material to significantly change the isotope ratio. In this case measurements of the change in isotope ratio due to cosmic ray exposure over a few thousand years would be meaningless given the contaminating aluminium and beryllium in the asteroid has been directly exposed to cosmic rays with no atmospheric attenuation for 4.5 billion years. If just 1 % of the asteroid were deposited on the landslide site, this would add 1 kg of aluminium and 2.83×10^{-6} kg of beryllium per square meter. It would also preferentially deposit material on the original surface as opposed to the surface exposed by the landslide and this could be the cause of the discrepancy between the two areas found by Kubik *et al*.

That NEO impacts can alter dating is suggested by some research at the Tunguska site by Franzén [56]. Samples for radiocarbon dating were taken from about 40 cm below the ground, which normally should have yielded a 14C-age of 300-500 years. However, the results showed that the samples produced a totally modern 14C-signal, or even a small "future" one [57]. In view of the meaningless results they were not formally published.

The cosmogenic radio nuclide surfacing dating technique is new and without Köfels has only one calibration point, that being the Younger Dryas. Further the Köfels results highlight that there are problems in obtaining conclusive dating as two sets of data have been found from the same site that differ by more than 2000 years, which is greater than the quoted two sigma error. It therefore is not a technique that can be regarded as providing definitive ages that can be used as disproof of hypotheses with alternative dates.

12.6 SECONDARY IMPACTS

Other impact sites are known to have associated secondary impacts from fragments that have broken off during the atmosphere flight. These secondary impact sites form in an ellipical pattern with the primary impact at the furthest point and the smallest fragments at the other end, the gradation due to the fact that smaller objects will be more influenced by aerodynamic drag [24].

Figure 38 shows a suggestive feature on the side of Felderkogel on the opposite side of the valley from Gamskogel (Figure 33). It has the form of a classic impact crater and it is in a position that is consistent with an object on a parallel path to the main object. The feature is around 250 m in diameter which would correspond to

Figure 38:   Felderkogel

an object a few tens of metres in diameter and in itself would have been a bigger impact event than the 1908 Tunguska impact.

12.7 SITE CONCLUSIONS

The Köfels site is possibly unique on the earth as the only obvious place where a significant geological feature has been created by a NEO exploding very close to the surface rather than either a high air burst or reaching the ground to form a crater. Without other examples for comparison it is difficult to establish what distinguishing features would either confirm or disprove Köfels as a NEO impact site.

It is clear that many of the features that led past researchers to propose it to be an impact site can now been seen as misleading and the result of the dynamics of a very large landslide – regardless of its cause. However there are aspects of the site noted by early researchers (notably Kranz [41]) that a simple landslide event cannot explain, and the site evidence remains consistent with a large close bolide as the cause of the landslide, and therefore consistent with it being the end point of the trajectory shown on Tablet K8538.

There are three sources for the current widely quoted age for the Köfels event at 8000 – 9000 years. Radiocarbon dating of embedded material has been seen as most secure and is used by the other two methods as a reference point. However there is no confirming data from post-slide material, and embedded material cannot have been on the surface when the asteroid exploded and the landslide occurred. With only embedded material the radiocarbon dating cannot be regarded as having any reliability, but assuming the buried trees were not contaminated with asteroid material, they may provide a maximum age, there is no radiocarbon evidence for either a minimum date or an actual date.

The fission track dating if taken in its own right suggests a Bronze Age date consistent with the Sumerian observation. The ^{10}Be and ^{26}Al nuclide dates can be stretched to make it possible to argue an Early Bronze Age date but not with any degree of conviction. However even within the data collected at the Köfels site there is sufficient variation in results that show the technique is not yet secure enough to produce unquestionable results and the possibility of surface contamination from the asteroid itself raises further questions about the accuracy of the technique for this site.

It is therefore concluded that the widely quoted age for the Köfels event at 8000 - 9000 years has no conclusive basis in published work. Within the data there is as much evidence for an Early Bronze Age date as for an Early Holocene date. So the site dating remains ambiguous and is not able to provide irrefutable proof one way or the other regarding the dates derived from Tablet K8538.

13

CONCLUSIONS

The tablet K8538 although clearly produced in the Assyrian period is a copy of an earlier Sumerian work. This is shown by the location of the celestial equator and the use of a pole star both of which preclude a First Millennium date and allows only a Sumerian period date. Other details suggest that the tablet was originally written in the earliest cuneiform.

The tablet presents a 360° of azimuth synoptic view of the night sky to an altitude of about 50° at a latitude of 31° – 33° just before sunrise at the end of June or early July in the Julian calendar, which corresponds to late May or early June in the Gregorian calendar. This is the only time when all the constellations shown by the tablet are simultaneously visible. Because it shows planets and clouds it is clearly a record of a specific night, rather than a general picture. The note style and some features strongly suggest it is an observation note made at the time by the observer.

Consideration of planets and the timing information on the tablet enable a time of 1:23 (Universal Time) on 29th June 3123 BC (Julian calendar) to be established. While this determination is not absolutely certain it is considered highly probable. More work on the Köfels site and in other areas such as ice cores may be able to confirm this date in a more conclusive manner.

The hypothesis that the tablet records the impact of the Köfels asteroid needs two steps to be proved. The first step is to show that a near Earth object (NEO) impact is recorded. This is the only possible interpretation of the text in the Pisces sector, which implies that the object there discussed was resolvable by the naked eye into finite proportions. This object could only have been asteroidal or cometary in nature. However the description of the body in the Path sector as "vigorously sweeping along" does not match the sedate apparent motion of a comet (normally requiring days for any discernible movement) and there is no doubt that a fast moving object of significant prominence is being described.

The second step is to determine that the NEO was the one that impacted at Köfels. Even with the rough track shown in the Pisces sector, Köfels is a strong contender as the impact site. With the precise trajectory recorded in the Path sector, with errors of less than 1°, there is no other known plausible candidate site for the impact.

These conclusions make K8538 a remarkable object that provides a view of a remarkable event that took place early in the morning on 29th June 3123 BC. A small group of astronomers performing a routine observation of the sky at a specialist observing site, recorded an incoming cloud front from the south east and the position of the planets.

Suddenly a bright object appears from behind the clouds, moving faster than anything they have ever seen in the heavens. They see it for around 4.5 minutes during which they make an excellent record of its track and appearance, which enables us 5000 years later to reconstruct the orbit and discover the object was an Aten object that had been subject to resonant capture by the Earth.

Five seconds before passing over the horizon from the location of the Sumerian observer the object would enter the Earth's shadow and become invisible. Although it is unlikely the observer realised this, to him it probably just appeared to go over the horizon. The object continued for 30 seconds when, over Greece, it entered the Earth's atmosphere and became visible first by its ionised shock front and then as a massive fireball with a sun-like intensity.

As the object travelled up the Adriatic Sea (inspiring pottery images in Hvar) and across the Alps the supersonic shock would have caused considerable destruction on the ground beneath the trajectory. The impact at Kofels, at 1.26 (Universal Time), would release energy equivalent to 1.4 x 10^10 tonnes TNT.

The Hvar pottery also has images of the massive plume. This plume would rise at an angle of 55 degrees to some 900 km before falling over the Levant and Sinai causing considerable destruction over a wide area. It would be visible from Sumer around 2 minutes after the object had disappeared over their horizon. The observer seems to have recorded this but regrettably that part of the tablet is missing. We are also missing the record of the aurora that has recently been rediscovered to accompany such large impacts from the Shoemaker Levy 9 impact on Jupiter.

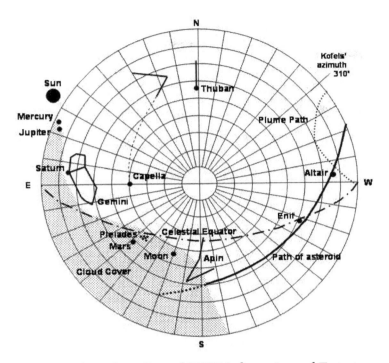

Figure 39: Planisphere View of K8538 Information and Trajectory

Figure 39 summarises the tablet information on Tablet K8538 assuming the 3123 BC date and the model trajectory on a true planisphere view.

The detailed knowledge of the trajectory provided by K8538 proves essential to finally resolving the many puzzling features of the Köfels' site. The absence of a classic impact crater is an inevitable consequence of the path over Gamskogel that caused the body to break up. The size of the landslide is due to the expansion of the 1 km class low density body over the 11 km from Gamskogel to Köfels.

The consistency of the evidence from the Planisphere, taken with that at the Köfels site, together with the derived resonant heliocentric orbit, all reinforce the conclusion that the tablet is a record of a direct observation of the asteroid that caused the Köfels feature.

This impact would not just be a spectacular sight. There would have been many direct casualties, near 100% mortality over areas of thousands of square kilometres in both the Alps and the Near East. There would also have been a severe global climate change that caused further death and social disruption. An event of this magnitude in the Early Bronze Age would be expected to leave some echo in the myths and records of the period. Clube and Napier [58] have considered such myths and show they are consistent with a major impact event in Europe in the Bronze Age. The event must have had a social, cultural and religious impact on those who experienced it and survived. It is therefore credible that an observation of the event would be preserved.

There are many ancient European and Asian myths that are not only generally indicative of a NEO impact, but contain detail very suggestive of this specific event. It was consideration of these myths that was the original focus of this study. While these myths do not constitute any form of proof that the Köfels impact happened, once the event is accepted, the detail they provide gives further insight into the processes involved with large NEO impact events.

When visiting the small alpine village of Köfels today (Figure 40) it is difficult to envisage the traumatic and immense event that happened 5000 years ago. The ground evidence for the impact, while there, is subtle. We are fortunate that our forebears left coherent and accurate accounts to help us unravel what happened.

Figure 40: Köfels Today

The observation made by the Sumerian astronomers in the Early Bronze Age was skilled, precise and objective and must be a candidate for the earliest known scientific work. The information it records is still of immense value over 5000 years later, not only because it gives an explanation of some of the history at the start of the Bronze Age and an insight into the level of human advancement at that time, but because it gives new understanding of the NEO capture processes, the dynamics of the entry and impact, and their global consequences.

REFERENCES

1. **Lugalansheigibar,** Untitled but commonly known as *"The Planisphere",* British Museum Catalogue No K8538 (3123BC)
2. **Bosanquet, R.H.M. & Sayce, A.H.,** "The Babylonian Astronomy No 2" *Monthly Notices of the Royal Astronomical Society,* Vol XL No 3, pp 105, (Jan 1880)
3. **King, L.W.** *"Cuneiform Texts from Babylonian Tablets in the British Museum"* Part XXXIII, (1912)
4. **Weidner, E.F.,** *"Handbuch der Babylonischen Astronomie"* (1915)
5. **Koch, J** *"Neue Untersuchungen zur Topographie des Babylonischen Fixsternhimmels",* Otto Harrassowitz, (1989)
6. **Hunger, H. & Pingree, D.,** *"Mul Apin, An Astronomical Compendium in Cuneiform"* Verlag Fredinand Berger & Sohne Gesellschaft M.B.H. (1989)
7. **Papke, W.** *"Die Sterne von Babylon"* Gustav Luebbe Verlag, Bergisch Gladbach, Germany (1989)
8. **Kramer, S.N.,** *"The Sumerians: their History Culture and Character"* University of Chicago Press (1963)
9. **Kramer, S.N.,** *"Sumerian Mythology"* revised edition, University of Pennsylvania Press (1961)
10. **Gassner, J-J.,** translated by Z. Bahrani & M. Van De Mieroop, *"The Invention of Cuneiform"* John Hopkins University Press (2003)
11. **Borger, R.,** Assyrisch-babylonische Zeichenliste, , Band 33 in Alter Orient und Altes Testament (AOAT), *Veröffentlichungen zur Kultur und Geschichte des Alten Orients und des Alten Testaments* (Series) Kevelaer and Neukirchen-Vluyn (1978)
12. **Bretagnon, P. & Francou, G.** "Planetary Theories in Rectangular and Spherical Variables. VSOP87 Solution" *Astronomy and Astrophysics* V202 pp309-315 (1988)
13. **Labat, R. &. Malbran-Labat, F.,** *"Manuel d'Epigraphie Akkadienne"* 6th edition, Librairie Orientaliste Paul Geuthner, S.A., Paris, ISBN 2-7053-0354-5 (1988)

14. **Lassine, A.** has placed a downloadable compendium of Borger's and Labat's sign list on the internet at www.alain.be/akkadien. Care needs to be taken as it does not use the conventional representation for the special Sumerian consonants. SH is used instead of š, KK instead of þ and NG instead of ñ.

15. **Halloran, J. A.,** Sumerian Lexicon version 3.0, downloadable from www.sumerian.org

16. **Brown, D.R.,** Mesopotamian Planetary Astronomy-Astrology, *Cuneiform monographs 18.* Groningen, Styx. (2000)

17. **Britton , J. & Walker, C.,** "Astronomy and Astrology in Mesopotamia" in C. Walker (ed) *"Astronomy Before the Telescope"*, British Museum Press (1996)

18. *"The Times Atlas of the World; Comprehensive Edition"*, 5th edition (1977)

19. **Richards, E. G.,** *Mapping Time,* Oxford University Press, ISBN 0 19 8504136 (1998)

20. **Crawford, H.** *"Sumer and the Sumerians"* 2nd edition Cambridge University Press (2004)

21. **Edwards, I.E.S., Gadd C.J. & Hammond N.G.L, (eds),** *"Cambridge Ancient History"* Vol 1 pt 2, 3rd edition Cambridge University Press (1971)

22. **Roaf, M.,** *"Cultural Atlas of Mesopotamia and the Ancient Near East"* Andromedia (1966)

23. **Petric, N.,** "Zanimljivi Cretezi Na Pretpovijesnoj Keramici Hvarske Kulture: Prikazi Kometa iz Priblizno 3000.G. Pr KR." *Vjesnik* 3series Vol. XXXV 11-18 (2002)

24. **Krinov, E.L.,** Giant Meteorites, Pergamon Press (1966)

25. **Dalfes, H.N.***et al* **(editors),** *"Third Millennium BC Abrupt Climate Change and Old World Order Collapse"* Springer-Verlag (1997)

26. **Bailie, M.,** *"Exodus to Arthur"* B.T.Batsford (1999)

27. **Brown, D. R.,** "The Cuneiform Conception of Celestial Space and Time" *Cambridge Archaeological Journal* 10:1, 103-121 (2000)

28. **Glasstone, S. &. Dolan, P.J.,** *"The Effects of Nuclear Weapons"* 3rd edition USDoD & USDoE, US Gov. Printing Office O-213-794 (1977)

29. **Ceplecha, Z. & McCrosky, R.E.,** "Fireball End Heights: A Diagnostic For the Structure of Meteoric Material", *Journal of Geophysical Research,* 81, pp 6257-6275, (Dec.1975)

30. **Sekanina, Z.,** "The Tunguska Event: No Cometary Signature in Evidence" *The Astronomical Journal,* 88, pp 1382-1414, (Sep. 1983)

31. **Melosh, H.J.** "Airblasts on Venus" *Nature* 358, pp622 - 623 (1992)

32. **Grieve R.A.F.,** "Terrestrial Impact: The Record in the Rocks" *Meteoritics* 26 pp175-194 (1991)

33. **Zahnla, K.,** "Tunguska: Leaving no Stone Unburned" *Nature* 383. pp674-675 (1996)
34. **Chyba, C.F.** et. al "The 1908 Tunguska Explosion: Atmospheric Disruption of a Stony Asteroid" *Nature* 361, pp40-44 (1993)
35. **Noll, K.S., Weaver, H.A.. & Feldman, P.D. (editors).** 'The Collision of Shoemaker-Levy9 and Jupiter', IAU Colloquium 156, ISBN 0-521-56192-2 (1995) – many contributions.
36. **Wing-Huen I.,** "Jovian Magnetospheric and Auroral Effects of the SL9 Impacts" in Noh, K.S., Weaver, H.A., Feldman, P.D., (editors)" *"The Collision of Comet Shoemaker-Levy 9 and Jupiter"* IAU Colloquium 156, Cambridge University Press (1995)
37. **Storzer, D., Horn, P. & Kleinmann, B.,** "The Age and The Origin of Köfels Structure, Austria", *Earth and Planetary Science Letters*, Vol 12 pp 238-244 (1971)
38. **Pichler, A.** "Zur Geognosie Tirols II, Die vulkanischen Reste von Köfels" *Jb.K.K.* Reichsant, Wien 13, (1863)
39. **Suess, F.,** "Der Meteorkrater von Köfels bei Umhausen im Oztztal, Tirol", *N jb Miner etc Beil.* – bd 72 Abt A, (1936)
40. **Stutzer, N.,**"Die Talweitung von Köfels im Otztal (Tirol) als Meteorkrater" *Z dt.geol ges* 88 523 (1936)
41. **Kranz, W.** "Beitrag zum Köfels-Problem: Die "Bergsturz-Hebungs – und Sprengtheorie", *Neues Jahrbuch fur Mineralogie, Geologie und Palaontologie abt B* Vol 80, (1938)
42. **Graham A.L., Bevan A.W.R. & Hutchinson R.,** *"Catalogue of Meteorites"* 4th edition British Museum (Natural History) ISBN 0-8165-0912-3 (1985)
43. **Shorts N.M. & Bunch, T.E.,** "A Worldwide Inventory of Features Characteristic of Rocks Associated with Presumed Meteorite Impact Structures" in French, B.M. and Shorts N.M. (editors) *"Shock Metamorphism of Natural Materials"* Mono Book Corp (1968)
44. **Heuberger, H., Masch, L., Preuss, E. & Schröcker, A.,** "Quaternary Landslides and Rock fusion in Central Nepal and in the Tyrolean Alps" *Mountain Research and Development 4*, 345-362 (1984)
45. **Masch, L., Wenk, H.R. & Preuss, E.,** "Electron Microscopy Study of Hyalomylonites – Evidence for Frictional Melting in Landslides" *Tectonophysics* 115 131-160 (1984)
46. **Erismann, T., Heuberger, H. & Preuss, E.,** "Der Bimsstein von Köfels (tirol) ein Bergsturz–'Friktionit'", in *Tschermaks Mineralogische und Petrographische Mitteilungen* 24 pp67-119, (1977)
47. **Lerouz, H. & Doukan, J.,** "Dynamic Deformation of Quartz in the Landslide of Köfels Austria" in *European Journal of Minerallogy* 5 pp893-902 (1993)

48. **Britt D. T. Yeomans D., Housen, K. & Consolmagno, G.,** "Asteroid Density, Porosity, and Structure", in William F. Bottke W.F. et al; *Asteroids III*, The University of Arizona Press (2002)
49. **Shorts N.M.,** "Experimental Microdeformation of Rock Materials by Shock Pressures from Laboratory-scale Impacts and Explosions" in French, B.M. and Shorts N.M. (eds) *"Shock Metamorphism of Natural Materials"* Mono Book Corp (1968)
50. **Surenian, R.,** "Shock Metamorphism in the Koefels Structure (Tyrol, Austria)" *Meteoritics,* Vol. 24, p.329 (1989)
51. **Lang, A., Moya, J., Corominas, J., Schroot, L. & Dikau, R.,** "Classic and New Dating Methods for Assessing the Temporal Occurrence of Mass Movements" in *Geomorphology* 30 pp33-52 (1999)
52. **Kubik, P.W., Ivy-Ochs, S., Masarik, J., Frank, M. & Schlüchter, C.,** "^{10}Be and ^{26}Al Production Rates Deduced from an Instantaneous Event Within the Dendro-calibration Curve", The Landside of Köfels, Otz Valley Austria *Earth and Planetary Science Letters* 161 pp 231-241 (1998)
53. **Kubik, P.W. & Ivy-Ochs, S.,** "A Re-evaluation of the 0-10 ka ^{10}Be Production Rate for Exposure Dating Obtained From The Köfels (Austria) Landslide" *Nuclear Instruments and Methods in Physics Research B* pp223-224 618-622 (2004)
54. **Lai, D.,** "Cosmic Ray Labeling of Erosion Surfaces: In Situ Nuclide Production Rates and Erosion Models" *Earth and Science Letters* 104 pp 424-439, (1991)
55. **Masarik, J. & Reedy, R.C.,** "Terrestrial Cosmogenic-nuclide Production Systematics Calculated from Numerical Simulations" *Earth and Planetary Science Letters* 136 pp 381-391 (1995)
56. **Jones, N.,** "Did Blast From Below Destroy Tunguska?" *New Scientist,* 7 page 14 (September 2002)
57. **Franzén. L.,** Private Communication (29th April 2005)
58. **Clube, V. & Napier, B.,** *"The Cosmic Serpent, A Catastrophist View of Earth History"* Faber and Faber, London (1982)

INDEX

16917508R00068

Made in the USA
Lexington, KY
19 August 2012